Dedication

For Rachel

ABOUT THE AUTHOR

John Henry is Professor Emeritus of the history of science at the University of Edinburgh. His interests lie in the history of interactions between science, medicine, magic and religion in the Renaissance and the early modern period. He is the author of *Moving Heaven and Earth: Copernicus and the Solar system* (2001, republished 2017), *Religion, Magic, and the Origins of Science in Early Modern England* (2012) and *The Scientific Revolution and the Origins of Modern Science*, 3rd edition (2008).

Knowledge is Power

How Magic, the Government and an Apocalyptic Vision Helped Francis Bacon to Create Modern Science

John Henry

9030 00005 7847 5

This edition published in the UK in 2017 by
Icon Books Ltd, Omnibus Business Centre,
39–41 North Road, London N7 9DP
email: info@iconbooks.com
www.iconbooks.com

Originally published in 2002 and 2003 by Icon Books Ltd

Sold in the UK, Europe and Asia by
Faber & Faber Ltd, Bloomsbury House,
74–77 Great Russell Street,
London WC1B 3DA or their agents

Distributed in the UK, Europe and Asia by
Grantham Book Services, Trent Road,
Grantham NG31 7XQ

Distributed in the USA by
Publishers Group West,
1700 Fourth Street, Berkeley, CA 94710

Distributed in Canada by
Publishers Group Canada,
76 Stafford Street, Unit 300,
Toronto, Ontario M6J 2S1

Distributed in Australia and New Zealand by
Allen & Unwin Pty Ltd, PO Box 8500,
83 Alexander Street,
Crows Nest, NSW 2065

Distributed in South Africa by
Jonathan Ball, Office B4, The District,
41 Sir Lowry Road, Woodstock 7925

ISBN: 978-1-7857-8236-7

Typesetting by Born Group

Printed and bound in the UK by Clays Ltd, St Ives plc

CONTENTS

LIST OF ILLUSTRATIONS

ACKNOWLEDGEMENTS

My interest in Francis Bacon was first kindled many years ago by Graham Rees and Julian Martin, two real Bacon experts, and I've gratefully drawn upon their work, as well as their inspiration, in the writing of this. More recently, I've been very grateful for long conversations with Silvia Manzo, an Argentinian scholar whose interest in Bacon testifies to his international reputation; it was good to be reassured, by another real Bacon expert, that my slant on Bacon wasn't too awry. I owe thanks also to Jon Turney, editor of this series, Simon Flynn at Icon Books, and John McEvoy, for excellent advice on how to make improvements.

I'd also like to take this opportunity to thank, for general encouragement and friendship over the years, Stuart McLeod, Andy Pearson and Mike Wardman.

Francis Bacon saw a wife and children as hostages to fortune, but mine have done nothing but improve my fortune. I'd like to thank my wife, Rachel, and my daughters, Eilidh and Isla, for their love and support, not just during the writing of this book, but always. Hoping there'll be future books that I can dedicate to my daughters, I dedicate this book to my wife, with especial thanks for everything, and with love.

• CHAPTER 1 •

'KINDLING A LIGHT IN NATURE'

Francis Bacon was a great genius who helped to shape the modern world. But many people would be hard put to say exactly why. He made no new discoveries, developed no technical innovations, uncovered no previously hidden laws of nature. His achievement was to offer an eloquent account of a philosophy and a method for doing those things. And in that way he turned out to be as important as people famed for particular discoveries, like Galileo or Isaac Newton, in what historians now call the Scientific Revolution.

Essentially, Bacon (1561–1626) wanted to reinvent investigation of the natural world. He was dazzled by a vision of progress whose ambition knew no bounds, stemming from a conviction as strong as that of a later philosophical revolutionary, Karl Marx, that the point of philosophy was not just to interpret the world, but to change it. As Bacon saw it in one of his soaring flights:

[A]bove all, if a man could succeed, not in striking out some particular invention, however useful, but in kindling a light in nature – a light which should

in its very rising touch and illuminate all the border-regions that confine upon the circle of our present knowledge; and so, spreading further and further should presently disclose and bring into sight all that is most hidden and secret in the world – that man (I thought) would be the benefactor indeed of the human race – the propagator of man's empire over the universe, the champion of liberty, the conqueror and subduer of necessities ('Proemium' (Preface), Of the Interpretation of Nature, *1603).*

In our terms, Bacon was a philosopher of science – perhaps the first one who really mattered. He was driven to combine three concerns: how knowledge was justified, how it could be expanded and how it could be made useful. His new method was designed to transform completely the knowledge of the natural world of his day, which he saw as both misconceived and sterile. As he was also a great writer, he helped to inspire others to adopt a new attitude to natural philosophy, an influence that lasted long after his death in 1626.

Two hundred and fifty years later, for instance, Charles Darwin described the method of working that was to lead him to his theory of natural selection. It was, he said, perfectly Baconian.

After I returned to England it appeared to me that ... by collecting all facts which bore in any way on the variation of animals and plants under domestication and nature, some light might perhaps be thrown on the whole subject. My first notebook was opened in July 1837. I worked on true Baconian principles, and without any theory collected facts on a wholesale scale ... (Autobiography, *1892, but written in 1876).*

Darwin didn't strictly work like this. Nor has any other successful scientist, because you cannot collect facts in this undirected way without disappearing beneath a mountain of irrelevancies. But the fact that the mild-mannered Victorian scientific revolutionary saw fit to invoke the Elizabethan statesman and philosopher is one index of Bacon's enduring fame. Another is the expectation we still hold, in a century when governments spend billions on research, that systematic experiment, conducted in a highly organised and institutionalised way, will yield useful answers to all kinds of problems – from acquired immunodeficiency syndrome (AIDS) to global warming. Our idea of science as the 'endless frontier', as the American Vannevar Bush titled a report to the US government just after World War Two, is also Baconian in spirit.

More recently, though, as increasing numbers of people question the impact of science, Bacon has attracted as many critics as admirers. Some deny his influence or significance. Others who acknowledge his importance see his attitudes as baleful or pernicious, his writings as a call to seek the domination of nature, and male domination at that. But all this scholarly argument just makes it more important to get to know what he was about. The fact that Bacon was concerned not with making scientific discoveries but with the very nature of science itself – seeking to establish what its purposes should be, the optimal methods for making discoveries, the best way to establish truth – has not only ensured his place in history but also ensured that it is controversial.

After all, while the value of a specific invention or discovery might be immediately recognisable, claims about the best way of going about things are endlessly debatable. When the subject for debate is something as culturally important as modern science and technology, it is hardly surprising that Bacon has continued to divide opinion. But what critics of Bacon sometimes fail to realise is that he himself was instrumental in making science and technology such characteristic aspects of Western culture. Before Bacon there was no such thing as science in our modern sense of the word. After Bacon, Western Europe was set on a course of discovery and invention that was to result in a civilisation based on the power of science and technology. In a very real sense, therefore, Bacon invented modern science.

In the text that follows we will look in some detail at Bacon's achievements, including why and how they came about. Before that, though, let's briefly consider just how Bacon can be said to be the inventor of modern science.

Chapter 2

Inventing Modern Science

Bacon's main claims to fame rest on three innovatory recommendations about the best way of acquiring knowledge of the natural world. These, in turn, led to other new attitudes to natural knowledge. The combination of Baconian methods with Baconian attitudes and expectations was to lead to the formation of something recognisably like modern science, and the modern scientific enterprise. What's more, it was to set science on the road to becoming one of the dominant aspects of modern Western culture.

To begin with, he was one of the very first natural philosophers (we have to call him that because there was no such thing as a 'scientist' in Bacon's day – just as there was no such thing as 'science' in our sense) to advocate the experimental method as the most efficient and reliable way of acquiring knowledge of the natural world. Before Bacon's time, the study of nature was based largely on armchair speculation. It relied almost entirely on abstract reasoning, starting from a restricted range of presuppositions about the nature of the world, and its aim was to explain known phenomena in ways that were consistent with those presuppositions. We now know that these presuppositions were incorrect and that much of pre-modern natural philosophy was therefore entirely misconceived, but this would never, could never, have been realised by anyone working within the tradition of natural philosophy (and therefore taking the presuppositions for granted). It needed someone like Bacon, willing to question the very foundations of natural philosophy, to ring the changes. Furthermore, there was no concern with discovery in traditional natural philosophy. When accidental discoveries were made, natural philosophers tried to accommodate them into their presupposed schemes. But there was no incentive to go out and make discoveries. Bacon helped to change all this by showing natural philosophers how successful and powerful the experimental method could be in discovering and understanding new facts about the world.

Second, Bacon believed that knowledge of nature should be turned to the benefit of mankind by exploiting new discoveries and inventions in a practical way. Familiar as this idea seems to us, it marked a radical break from the speculative and contemplative study of nature that was typical in Bacon's day, and that was concerned with understanding for its own sake. Bacon's insistence on putting knowledge to use was to have the most far-reaching consequences, not only in stimulating specific discoveries and inventions, but also in changing everybody's attitude towards, and expectations of, scientific knowledge.

Third, Bacon had an equally innovatory view of what has been called 'the logic of scientific discovery'. The prevailing rationalist natural philosophy upheld deductive logic as the only sound form of reasoning, but Bacon insisted that deductive logic was only useful for confirming what was already known, rather than for pointing to new discoveries. Since he believed that what was already 'known' by traditional natural philosophy was mostly wrong, Bacon went so far as to say that deductive logic perpetuated error by confirming as true what was in fact false. As an alternative, he suggested a refinement of what is known as 'inductive logic', the logic of everyday experience.

If we always suffer a headache after drinking red wine, we might suppose that red wine gives us a headache. In doing so we are instinctively deploying inductive logic. All too often, though, this is not a reliable form of logic. Maybe our headaches are caused by something else we do whenever we drink red wine, but that we haven't noticed. Accordingly, Bacon tried to develop an elaborate procedure that would, he believed, make induction more certain. This last aspect of Bacon's philosophical reforms has been the most controversial and has divided philosophical opinion about its validity. It was nevertheless very influential in the subsequent history of science.

These three recommended changes in our approach to the understanding of nature were the primary features of Bacon's teachings, but they led to important secondary features, which are now no less characteristic of modern science.

Science is now inseparably linked with notions of progress. As scientific knowledge advances, so does our way of life in the West. But this, too, is part of the Baconian legacy. Before Bacon, the general assumption among intellectuals was that knowledge must be recovered from the past. Adam, the first man, had known all things before the Fall and the expulsion from the Garden of Eden (see 'Glossary' for brief explanations of the terms and religious doctrines that appear throughout this book). Since then, wisdom had been progressively forgotten. Nobody thought that the truth of things was waiting to be discovered at some future time; it must be recovered from Ancient writers who lived nearer to Adam's time, and who had forgotten less.

When Nicolaus Copernicus suggested in 1543 that the Earth went around the Sun, instead of the Sun around the Earth, his followers referred to it as the Pythagorean theory, not the Copernican theory. If the theory was true, the evidence for it must lie in past wisdom and, sure enough, they discovered

Ancient Pythagoreans who had reputedly said the Earth moved. But Bacon was to change this – not overnight and not single-handedly, but he was the first to point the way forward to the future.

The sciences have made 'but little progress', Bacon wrote, because '[i]t is not possible to run a course aright when the goal itself has not been rightly placed'. Bacon showed the way: 'Now the true and lawful goal of the sciences is none other than this: that human life be endowed with new discoveries and powers' *(New Organon*, or *Novum Organum*, I, Aphorism 81 – see Chapter 4 for a further explanation of this work). Nobody before Bacon urged that scientific knowledge should be put to use for improving mankind's lot. Once Bacon put this idea forward, it led to expectations that scientific knowledge should lead to progress, not just in our knowledge, but through that to an improvement in our society.

Modern science is also regarded as the supreme form of objective knowledge. Where other claims can always be regarded with suspicion or merely a healthy scepticism, scientific claims can at least be seen as being capable of substantiation in ways that are free from any cultural, ideological or personal bias. The objectivity of scientific knowledge has come under increasing attack in recent years, particularly by sociologists of science who seek to show that it, too, is culturally biased. We needn't go into these arguments here, suffice it to say that *relatively speaking* scientific knowledge remains the most reliable and objective form of knowledge we have. (Compare a scientist's knowledge claims, for example, with those of a politician, a religious leader, a literary critic, a pundit speaking on behalf of modern art or anyone who believes, like the nineteenth-century Romantic poet John Keats, that the arts can speak truth. 'Beauty is truth, truth beauty', Keats wrote in his 'Ode on a Grecian Urn', but it would be foolish to take him literally.)

This supposed objectivity of scientific knowledge, in so far as it does exist, is also the outcome of Bacon's views on science. The concept of objectivity had not been thought of in Bacon's day. Things were thought of simply as being true or not true. But once, under Bacon's influence, scientists began to think about the best way of establishing truth, the notion of objectivity emerged. Whether objective knowledge is really possible or not (and sociologists would say it isn't), it is clearly better to *aspire* to a knowledge that is free from ideological bias rather than to promote claims to truth that have been deliberately conceived to support a particular ideology or ungrounded system of belief. The aspiration to objectivity in science is another Baconian legacy.

Given that modern Western science is characterised as progressive and objective, and that both these aspects are seen as being linked to the dominant presumption that scientific knowledge should be practically useful, it is impossible to deny, despite what critics might say, that the ethos and practice of modern science are essentially Baconian. Similarly, even the massive and complex organisation of modern science can be seen as the fulfilment of a Baconian dream.

Francis Bacon was a career civil servant who rose to be Lord Chancellor of England, and he always believed that the fully comprehensive and practically useful science he envisaged could only properly be pursued under the aegis of the state. The modern science-and-technology complex of advanced societies does indeed depend almost entirely on state funding and support, and many scientists are civil servants or, as in the case of research scientists in government-funded universities, effectively servants of the state. If Bacon himself never succeeded in having his plans for reform of scientific knowledge taken up by the government of which he was a part, his beliefs about the need for state support of science have certainly proved to be prophetic.

It can hardly be denied that there is much in modern science – whether we are talking about its intellectual content, its methods or its social and institutional organisation – that was never envisaged by Bacon. But no matter how far it has developed beyond anything that Bacon could have predicted, it has never lost sight of its Baconian beginnings in an objective experimental method allied to a concern to develop practical benefits leading to intellectual and social progress. Modern Western science remains a Baconian enterprise. We will look more closely at each of these aspects of his work in what follows, but before that we should consider where these ideas came from. Why was it that Bacon, and nobody else, developed these revolutionary innovations in the method of doing science?

There is no denying that Bacon was a great genius who helped to shape the modern world, but geniuses do not spring from nowhere. We want to know where their ideas might have come from, why they thought differently from others (but not so differently that nobody else could see the point of their innovations), why they were so passionate to change what others were complacent about. The historian is like a private investigator, not content to reveal that a suspect had the opportunity to commit the crime, but seeking also to strengthen the case by uncovering the motivation.

In the mysterious case of Francis Bacon, we will see that the story of how he came to do what he did takes us through unexpected and unfamiliar territory – through realms that now seem to have little or nothing to do with natural science, but that were once intimately connected with it. These are the realms of religion and magic. In the twenty-first century, religious and magical beliefs seem to be completely incompatible with, and even anti-thetical to, scientific knowledge. But there is a history behind the separation of these things, and Baconianism itself is part of that history.

There was a time, however, when religion, magic and knowledge of nature were much more closely connected, and even interdependent. So much so that religion and magic played their parts in moti-vating the life's work of Francis Bacon.

CHAPTER 3

'BELIEVING THAT I WAS BORN FOR THE SERVICE OF MANKIND'

If we want to understand Bacon's motivation we must look for clues to his psychology. By searching through his writings we can find, here and there, comments in which he seems to reveal his innermost thoughts and beliefs. One of the most famous examples of such revelations occurs in a preface that Bacon wrote, in 1603, for a work he later abandoned. Here Bacon tells us how he mused about his destiny:

Believing that I was born for the service of mankind, and regarding the care of the common-wealth as a kind of common property which like the air and the water belongs to everybody, I set myself to consider in what way mankind might best be served, and what service I was myself best fitted by nature to perform ('Proemium', Of the Interpretation of Nature).

Bacon's belief that he was 'born for the service of mankind' was not merely vanity and conceit. Though this might seem like an odd comment if it had been made by virtually anyone else, it is hardly surprising that Bacon should have thought this way about himself. He was in fact deliberately raised by his father, Sir Nicholas Bacon, to play a part in the high offices of the state. Sir Nicholas was himself Keeper of the Great Seal of England, making him one of the chief ministers under his sovereign, Elizabeth I, and he continually primed Francis and his older brother Anthony for service in the Elizabethan common-wealth. Although Francis talks in the quotation above about service to mankind in general, he shortly after acknowledges that, because of his family and educa-tion, he tended to think primarily of his own country.

So, what conclusion did Bacon reach about the service to mankind he was most suited to perform? Obviously he wanted to improve things somehow, but he was all too aware that works in politics – founding cities, establishing laws, deposing tyrants and the like – tended to be restricted in place and time. What was much more beneficial in the long run, he believed, was 'the work of the Inventor': 'Now, among all the benefits that could be conferred upon mankind, I found none so great as the discovery of new arts, endowments, and commodities for the bettering of man's life' ('Proemium').

Elsewhere in his writings, Bacon tells us of the three great inventions that undoubtedly made him think along these lines – the printing press, gunpowder and the magnetic compass:

For these three have changed the whole face and state of things throughout the world; the first in learning, the second in warfare, the third in navigation; whence have followed innumerable changes; insomuch that no empire, no sect, no star seems to have exerted greater power and influence in human affairs than these changes (New Organon, *I, Aphorism 129*).

Figure 1: Francis Bacon, Viscount St Alban (1561–1626) – philosopher, Lord Chancellor.

It may seem like setting one's sights too high to decide to be not just a run-of-the-mill inventor but one trying to come up with something likely to have as big an impact in the world as printing, gunpower or the compass. But, in fact, Bacon set his sights even higher. As we have already noted, Bacon wanted to kindle 'a light in nature', not by 'striking out some particular invention' but by inventing a method, a set of procedures, that would enable mankind to 'disclose and bring into sight all that is most hidden and secret in the world'. This is Bacon's wonderful image of his life's calling. Unconcerned about making a particular discovery or invention, Bacon saw himself as 'the propagator of man's empire over the universe, the champion of liberty' ('Proemium'), lighting a way for others to investigate nature to best effect and so make discovery after discovery, until perhaps everything was known.

What's more, Bacon saw this as something that would benefit all humankind. Here was an ambition that transcended anything his father had taught him about being a useful servant to his country. Bacon discerned three grades of ambition:

The first is of those who desire to extend their own power in their native country; which kind is vulgar and degenerate. The second is of those who labour to extend the power of their country and its dominion among men. This certainly has more dignity, though not less covetousness. But if a man endeavour to establish and extend the power and dominion of the human race itself over the universe, his ambition ... is without doubt both a more wholesome thing and a more noble than the other two (New Organon, *I*, Aphorism 129).

Immediately, this raises a question. Why did Bacon go beyond his father's teaching and expectations? As far as we can tell, Sir Nicholas was concerned only with the second grade of ambition, and there is nothing to suggest that he might have inspired his younger son to set his ambitions higher. As Bacon himself wrote in his autobiographical preface: '[M]y birth and education had seasoned me in business of state; and ... I thought that a man's own country has some special claims upon him.' So what might have led Francis to rise above this to the third grade of ambition? For the answer to this question we should perhaps turn to Bacon's mother.

Anne, Lady Nicholas Bacon, was one of Sir Anthony Cook's three daughters, all of whom were famed in their day for their learning (at a time when women were usually denied any formal education). Her father had been tutor to the young King Edward VI (and no doubt had a hand in the boy sovereign's tendency to Calvinism), and the direction in which Anne's learning took her is indicated by the fact that she quickly published her English translation of Bishop John Jewel's *Defence of the Church of England* (published in Latin in 1562) when Francis was just three years old.

Jewel was one of the principal architects of English Protestantism, playing an important role in the establishment of Anglicanism after the brief Catholic reign of 'Bloody Mary' (Mary I, 1553–58), and his Defence was a major statement of reformed religion against the corruption of the Catholic Church. Anne Bacon, as her speedy translation of Jewel's work shows, was one of the earliest and staunchest Protestants in Elizabethan England. A Calvinist in her theology and Puritan in her morality, she introduced family worship and Bible reading into her home, and there can be no doubt that her religion had a significant influence on her young sons.

For those of us living where Western culture is dominant, we are in a secular age. In our society, belief in God is a matter of personal choice and an entirely private matter, but it was very, very different in Bacon's day. It is hard now for us to imagine a society in which Christianity was a characterising aspect of the culture, much less to understand how that would have affected day-to-day thinking.

For Bacon and his contemporaries, God and religion were so pervasive in social, political and intellectual life that, to a large extent, systematic disbelief was practically impossible. As far as we can tell, there was no such thing as an atheist before the late sixteenth century (heretics yes, but atheists no), and even in the seventeenth century their existence is dubious and, at most, extremely rare.

For us, atheism is a perfectly rational position to adopt (indeed, atheists tend to think of it as the only rational position), but in Bacon's day atheism was barely comprehensible in a world that was thoroughly suffused by religion and where God seemed to be a continual presence. Disbelief in religion then would have been as baffling as if someone today said they did not believe in science – it would be difficult for us to know what somebody could possibly mean by that.

So, for early modern thinkers, and especially for those (like Bacon) raised in a thoughtfully devout household, the religious perspective provided the only possible point of view. There might be disputes over the fine details of doctrinal theology that separated Protestants from Catholics (or from other kinds of Protestant), but there was no questioning the fundamental rightness of the religious world view. God was in His heaven and all was right with the world. And there are strong indications that Bacon saw his great enterprise – his wish to benefit not just his own countrymen but all mankind – as part of his Christian duty.

When Bacon finally published his so-called *Great Instauration* (1620), which in fact was nothing more than a summary outline and a couple of sample parts of what the real 'Great Instauration' – the name he gave to his ambitious reform of natural knowledge – was meant to be, he provided it with an evocative frontispiece (see Figure 2). At the bottom of the picture Bacon included what might be regarded as his favourite motto: 'Many will go to and fro, and knowledge will be increased' (*Multi pertransibunt et augebitur scientia*). This is a quotation from The Book of Daniel (12, 4). Daniel is to the Old Testament what The Revelation of St John, often known simply as Revelations, is to the New Testament: a prophetic book that describes what will happen before the world comes to an end. At this point the reader is being told that completion of the times, and the end of history, will be presaged by the fact that many will go back and forth and that knowledge will increase.

According to Bacon, the first part of this prophecy had already come to pass. Many going to and fro was a reference to the Renaissance voyages of discovery and the subsequent opening up of new trade routes around the world. Many were now circumnavigating the globe, and passing to and fro across the Atlantic, or around the coast of Africa into the Indian Ocean. All that remains, therefore, is for science to advance. 'What else can the prophet mean ... in speaking about the last times?' Bacon asks in his *Refutation of Philosophies* (1608). 'Does he not imply that the passing to and fro or perambulation of the round earth and the increase or multiplication of science were destined to the same age and century?'

So if Bacon's system of reforming natural philosophy works, then science will advance immeasurably (he believed), and the world as we know it will come to an end as the Second Coming of Christ takes place on the Great Day of Judgement. For the unfaithful, of course, the end of the world could only be an unmitigated disaster (to put it mildly). But for religious believers, at least those who were confident of being among the chosen rather than among the damned, it was a consummation devoutly to be wished.

It was no doubt the same kind of religious thinking that led Bacon to entitle one of the parts of the *Great Instauration 'Parasceve ad historiam naturalem et experimentalem'*. Usually translated simply as 'Preparation for a natural and experimental history', this fails to pick out the significance of Bacon's use of the special word *'parasceve'*. It is a Greek word, not a Latin one, but it would have been familiar to Bacon's educated readers as the word used in the Greek New Testament to refer to the day of preparation for the Sabbath – the day when all tasks have to be completed, so that no work needs to be done on the Day of Rest. Here again, the Sabbath that Bacon now had in mind was not just the weekly Sabbath, but the ultimate Sabbath of the Day of Judgement. His natural and experimental histories, therefore, were to be written in partial fulfilment of Daniel's prophecy, a collection of the latest scientific knowledge to demonstrate the recent advance of science.

Bacon referred to the Last Day more explicitly at the end of his outline 'Plan of the Work' for the *Great Instauration*. Having criticised earlier philosophers for mentally concocting elaborate theories about the nature of the universe without checking them against physical reality, he wrote:

Figure 2 (opposite): The frontispiece to Bacon's *Great Instauration* (1620). It depicts a ship passing through the pillars of Hercules, the supposed limits of the Ancient world, into the ocean beyond. The intention was not only to remind the viewer of the recent voyages of discovery that had expanded the known world, but also to suggest the voyages of intellectual discovery that would be possible by pursuing Bacon's new method. The legend at the bottom, a quotation from the Book of Daniel in the Old Testament, associated voyages of discovery with the advancement of the sciences in a prediction of the end of the world.

FRANC. BACONIS
DE VERULAMIO/
Summi Angliæ
CANCELLARIJ/
Novum Organum
Scientiarum.

Multi pertransibunt & augebitur scientia

LVGD. BAT.
Apud Adrianum Wijngaerde.
et Franciscum Moiardum .1645.

God forbid that we should give out a dream of our own imagination for a pattern of the world: rather may He graciously grant to us to write an apocalypse or true vision of the footsteps of the Creator imprinted on His creatures ... Wherefore if we labour in thy works with the sweat of our brows thou wilt make us partakers of thy vision and thy Sabbath.

Raised by his father to be a public servant, Bacon clearly saw it as his destiny to do something *pro bono publico* (for the good of the public). He might have been content here to serve the English public, or perhaps we should say the British public, since Bacon achieved high public office not under Elizabeth but under her successor James VI of Scotland, who became, after the union of the Scottish and English crowns in 1603, James I of England. But Bacon wanted to see himself as a benefactor to all mankind, an ambition he saw as more wholesome and noble, and which seems to have been inspired by his devotion to his Protestant faith with its belief, widely held at that time, that the end of the world was not far off.

It is also clear that Bacon believed that a reformation of the system of natural knowledge, including a new method of discovery and of confirmation, was of more benefit to mankind than any political changes he could realistically make. Furthermore, this intellectual enterprise, unlike other political manoeuvring, was sanctioned by Holy Writ as a means of fulfilling Biblical prophecy about the advance of science ushering in the Day of Judgement.

These, then, were the roots of Bacon's ambitious scheme to reform knowledge. And they show to some extent why his suggested reforms took the shape they did, emphasising a practical usefulness for the amelioration of the human condition. But other aspects of his new method of science remain unexplained. Why, for example, did Bacon insist on the use of experimental investigation? The experimental method was by no means routinely in use in natural philosophy before Bacon's time, and Bacon undeniably played a major role in introducing experimentalism into natural philosophy. But he did not invent the experimental method. He borrowed it. Furthermore, as we shall see, he borrowed it from an old tradition that had already been concerned to improve the human condition, albeit on a more individualistic scale, for many centuries before Bacon came along. We'll come to this soon, but first, let's take a brief look at Bacon's life and times.

• CHAPTER 4 •

'BEING OF ALL MEN OF MY TIME THE MOST BUSIED IN AFFAIRS OF STATE'

Bacon is often referred to as the great statesman of science, and he was, in his own lifetime, first and foremost a statesman. What's more, the specific nature of his scientific philosophy owed a great deal to the fact that he was a civil servant, being shaped in significant ways by bureaucratic concerns that must have been second-nature to Bacon (see Chapter 11).

As we've already seen, he was put on to his life's course by his father, Sir Nicholas Bacon, Elizabeth I's Lord Keeper of the Great Seal, who began by sending Francis to Trinity College, Cambridge, when he was twelve years old. This was younger than usual but not unheard of; universities were a bit different in those days and Bacon was not studying for a degree like a modern student.

He left after only two years' residence in Cambridge (interrupted by periods of plague), and entered Gray's Inn in London in 1576 for training in English common law. Almost immediately, though, he went to live in France to learn about the practice of Roman Law, which held sway throughout the rest of Europe. He was in Paris in 1579 when his father died unexpectedly and, as far as his family were concerned, with unexpectedly large debts. The will left no special provision for Francis, merely an injunction on Anne to look after him and his brother. However, it did make generous provision for the family of Sir Nicholas's first wife. To say the least, this was a set-back in Bacon's fortunes.

Francis continued with his legal studies at Gray's Inn, but he also embarked on a long career of searching for 'preferment' in some government post, which always entailed finding a suitable patron (then, as now, it was not *what* you knew that counted, but *who* you knew). In this regard Bacon was pretty well placed. William Cecil, Lord Burghley, was Bacon's uncle and Queen Elizabeth I's chief secretary of state, and Bacon was already importuning him for help up the career ladder in 1580. On other occasions he sought the patronage of Sir Francis Walsingham and Robert Dudley, Earl of Leicester. However, his public career began when he was elected to the House of Commons in 1581. He proved to be a good parliamentarian, being both knowledgeable and highly articulate in debate, and he continued to be returned to Parliament until his public disgrace in 1621 (which we will learn more of later). Meanwhile, he also built up a considerable reputation as a lawyer.

Bacon's first real successes on the road to government office came when Robert Devereux, 2nd Earl of Essex and a rising favourite of Elizabeth's, became his patron. Bacon was now appointed to a number of committees, and short political tracts that he wrote supporting establishment policies began to be circulated in manuscript in government circles. In 1589 he was granted the office of Clerk of the Council in the Star Chamber, but he had to wait until the current incumbent vacated the post. This would have earned Bacon £1,600 a year, but he had to wait nearly 20 years before the post was finally vacated for him.

This promise was the best that Bacon was ever to achieve under Elizabeth. He made the mistake in 1593 – at least it was a mistake from the point of view of his career ambitions – of opposing the Crown's request to the Commons for extra subsidies. This showed Bacon to be a good parliamentarian, but it enraged Elizabeth and evidently set her against him. When he applied for the vacant post of attorney general the Queen offered it to the current solicitor general; and when Bacon then applied for the vacated post of solicitor general, the Queen still managed to think of someone more suitable.

If any good came out of this for Bacon, it was to be found in the fact that he now recognised that Essex was perhaps not the best patron.

Consequently, he began to distance himself from him. Even so, it seems that Bacon could only fully avoid being implicated in Essex's subsequent treason by acting as the prosecutor of his former friend. Essex finally fell from grace with Elizabeth (they'd already quarrelled about the way to deal with the Irish rebellion, and Elizabeth had struck Essex in public) when he negotiated a humiliating truce with the Earl of Tyrone in 1599, after failing, at the head of a large army, to crush Tyrone's rebellion in Ireland.

On returning to England, Essex was imprisoned and suspended from all his offices. After his release some months later, he was banished from the court and was now able to brood on his failures. Being repeatedly refused access to the Queen, he became convinced that Elizabeth's advisers were plotting against him and he responded with a more dangerous plot of his own – to seize the Queen, and force her to dismiss her advisers and make changes in the state. Essex gathered numerous malcontents about him and, after being disturbed by Elizabeth's agents sent to investigate the unlawful assembly, he marched into London, hoping to inspire a popular revolt.

The subsequent trial was incompetently prosecuted by Sir Edward Coke, the attorney-general, and it was Bacon who had to clarify and press home the case against Essex. It's been said that Bacon's arguments played the major role in establishing the guilty verdict. This was in spite of the fact that part of Essex's defence was to insist that he was acting on the advice of Bacon. It's certainly true that Bacon had continually offered advice to Essex since before his expedition to Ireland, but Essex never quite took it the right way, and went on to do much that Bacon would never have condoned. It was perfectly fair of Bacon to claim at the trial, which was held in Westminster Hall on 19 February 1601, that 'I have spent more time in vain in studying how to make the earl a good servant to the Queen, than I have done in anything else'.

Even so, after the succession of James I (1604), Bacon responded to 'common speech' against his behaviour by publishing an apology for his involvement in the prosecution of his former patron and friend. Bacon's defence of his actions was the only one possible: that he put his duty and loyalty to the Queen above his friendship. Indeed, he wrote:

For every honest man, that hath his heart well planted will forsake his King rather than forsake God, and forsake his friend rather than forsake his King; and yet will forsake any earthly commodity, yea and his own life in some cases, rather than forsake his friend (Apology in Certain Imputations Concerning the Late Earl of Essex, 1604).

These are fine words, but they glibly overlook the fact that Essex clearly did not intend to depose or harm the Queen and that, had Bacon let the main prosecutor, Sir Edward Coke, do all the talking, his incompetence might have told in Essex's favour. Samuel Johnson once said that patriotism was the last refuge of a scoundrel. It seems hard not to think of Bacon's claims of loyalty to the crown to be a similar refuge or, worse, another attempt to win favour and preferment from the Crown even at the expense of his friend's life.

One way or another, Elizabeth's reign was a series of disappointments for Bacon. Not only had he failed to achieve any official place in affairs of state, but his ambitions to reform natural philosophy and to kindle a light in nature for the benefit of mankind, or even for the commonwealth of England, received no encouragement whatsoever. Under her successor, James I, things certainly improved with regard to state preferments. Bacon was knighted in 1603 and further noticed by James for his parliamentarian work in connection with the union of the Crowns of Scotland and England (James was already King of Scotland). Bacon finally became solicitor general in 1607.

The following year he was also able, at last, to take up the post and the salary of the Clerk of the Council in the Star Chamber. His fortunes improved dramatically after the death of Robert

Cecil, the Earl of Salisbury, Lord Burghley's son, who like his father rose to be Elizabeth's secretary of state and who became James's chief minister. The Cecils had formed a rival faction at court in the days when Essex was the Queen's favourite, and Cecil continued to hinder Bacon's advancement until his death. But Bacon did not rely merely on the removal of an obstacle. He also ingratiated himself with the King's rising favourite, Sir George Villiers, later Duke of Buckingham.

In 1613 he became attorney general and in 1616 a member of the King's Privy Council. When Villiers secured for Bacon James's promise that he would be next Lord Chancellor, Bacon wrote in gratitude to him: 'I am yours surer to you than my own life ... I will break into twenty pieces before you have the least fall.' It's a good job that Bacon never found himself in court having to prosecute him.

Bacon achieved the same position as his father in 1617, Lord Keeper of the Great Seal and, as promised, was appointed Lord Chancellor the following year. He also now became Lord Verulam (the Latin name for St Albans), and Viscount St Albans in 1621. He'd reached the top and he had to stop anyway, but it was at this point that he took a serious tumble.

A campaign was raised against him from the House of Commons in 1621 accusing him of accepting bribes from plaintiffs in his court cases.

Bacon was surely right to point to the 'vices of the times' in partial justification of what he did. Certainly it was not unusual then for gifts to change hands in cases of litigation. But Bacon seems to have been overstepping even the usual vague marks and taking things too far – although in many cases, it seems it was members of his household who were taking the bribes, without telling Bacon.

He confessed, anyway, to having done enough to deserve condemnation and censure. He was sentenced to the Tower of London during the King's pleasure, fined £40,000 and forbidden to hold state office or to sit as a Member of Parliament. The King was able to intercede on Bacon's behalf. He stayed in the Tower for just a few days and the fine was remitted. The worst punishment was the loss of his earnings which, among other things, forced him to sell York House, the London residence of his family, to the covetous Buckingham.

And so it was that Bacon was forced to give up all affairs of state in the last years of his life. At last, he could devote his time to the enterprise he always believed would bring him a more lasting fame. Bacon now seemed to regret the lost time spent in seeking and fulfilling high office, and wished he had devoted his time to his plans to reform natural knowledge. Shortly after confessing to bribery and corruption to the House of Lords, he wrote a different confession to his God:

Besides my innumerable sins, I confess before thee, that I am debtor to thee for the gracious talent of thy gift and graces, which I have neither put into a napkin, nor put it (as I thought) to exchangers, where it might have made best profit; but misspent it in things for which I was least fit; so that I may truly say, my soul hath been a stranger in the course of my pilgrimage (Prayer After Making His Last Will, *1625).*

Bacon had begun writing pieces on the need for, and the way to go about, reform of natural philosophy in 1603, but it was only in 1620 that he published a description and outline of what he now called his Great Instauration. Bacon told King James that he had been working on the book for 'near 30 years', which, judging from short pieces written in the 1590s that foreshadow his ideas for reform of natural philosophy, seems about right. The Instauration was envisaged in six parts:

- The Divisions of the Sciences
- The New Organon, or Directions concerning the Interpretation of Nature
- The Phenomena of the Universe, or a Natural and Experimental History for the foundation of Philosophy
- The Ladder of the Intellect

- The Forerunners, or Anticipations of the New Philosophy
- The New Philosophy, or Active Science.

The first of these was simply intended to be a summary of the knowledge already possessed by the human race – a kind of 'before' to compare with Bacon's 'after'. The second part was concerned with describing the precise method of science, the procedures to be used for establishing and interpreting our factual knowledge of the natural world with a view to arriving at general axioms, or laws of nature. The third part was intended to provide the repository of factual knowledge upon which the Baconian method was brought to bear. The second and third parts of Bacon's Great Instauration went hand in hand.

The method demanded, and depended on, as comprehensive a collection of relevant information as was humanly possible; the third part was dedicated to gathering, compiling and sorting that information on a grand scale. It was to be a compendium or encyclopaedia of natural phenomena, and of human skills, techniques and crafts that depend on those phenomena. Bacon believed that the required collection of comprehensive natural and experimental histories, as he called them, necessitated a massive collaborative effort and would take many generations to compile.

The end result of the Great Instauration, therefore, would inevitably be delayed.

The 'New Philosophy, or Active Science' of part 6 was to be the outcome of applying the method in part 2 to the data in part 3. Although Bacon believed it would be a mistake to try to define part 6 before all the possible data of part 3 had been collected, in some restricted cases, he hoped, it might be possible to illustrate the kind of thing to expect. This is how Bacon saw parts 4 and 5: preliminary examples of how the method of part 2 should be applied to the data in part 3, and interim examples of useful discoveries to serve as indications of the eventual benefits to arise from the Great Instauration.

Until it achieved completion, then, with the 'New Philosophy, or Active Science', the Great Instauration essentially depended on the development of a precise method for the gathering, analysing and correct understanding of information about the natural world, and the collection of a huge database, a compendium, catalogue or encyclopaedia of all known phenomena, to provide the raw data upon which the method could be brought to bear.

In the 1620 publication, under the title *Great Instauration*, Bacon included a major indication of the methodological part (part 2) of the enterprise. This was the *New Organon*, or New Instrument, which took its name from the Organon, or Instrument, of the prevailing authority in all things philosophical, the Ancient Greek philosopher, Aristotle (384–322 BC). (*The Organon* was not the title of a book by Aristotle but the collective name for his several books on logic and method used in the universities at that time.) This turned out to be the only part of the *Great Instauration* that Bacon was able to develop in anything like a complete form, and even this remains unfinished. Nevertheless, it is considered the most substantial part of his output in natural philosophy, and it is the work upon which his reputation as a philosopher chiefly rests. The other main part of the enterprise, the encyclopaedic database, was illustrated at this time by the *Parasceve* (remember, it literally means 'preparation for the Sabbath'), the 'Preparation for a natural and experimental history'. But this was just one of several efforts Bacon made to single-handedly foreshadow the kind of compendium that could only properly be achieved collectively and over many years.

After his impeachment, for example, Bacon published another sample of the kind of thing he had in mind for the encyclopaedic part of his Great Instauration. This was the *Natural and Experimental History for Establishing Philosophy* (1622). Consisting chiefly of a collection of information on winds, the *Historia Ventorum*, it also included yet another prospectus of what was to come. It is an indication of Bacon's determination to make up for the lost time of his years as a mere statesman that he announced here his intention to write six more specimen 'histories', or collections of data (he uses the word 'history' as it appears in the phrase 'natural history'; he doesn't mean an account of past events), over a period of six months.

In the event he managed to publish the *History of Life and Death* in 1623, and wrote but left unpublished the *History of Density and Rarity*. The fact that he failed to produce the rest of these projected histories at the rate of one a month does not mean that Bacon ground to a halt. On the contrary, in these final years of his life he wrote several other major works. There was, for example, the *New Abecedarium of Nature* (1622), which he saw as an indication of how to apply the methods and procedures described in the methodological part of the Great Instauration (in the *New Organon*) to the information in the natural histories. (Incidentally, by 'Abecedarium', which should be pronounced like reciting the alphabet, A B C-darium, Bacon means to imply a survey of nature from A to Z – that is, covering everything.)

He also published one of his most important works, the much-expanded Latin version of *The Proficiency and Advancement of Learning* (1623, built on a shorter English version published in 1605), and wrote the *Sylva Sylvarum* and *The New Atlantis*, two of his most important works (published shortly after his death). He also began, but left unfinished, his *History of Animate and Inanimate, An Inquiry Concerning the Lodestone*, and *Topics Concerning Light and Brightness*. What's more, these are only the scientific works. Between his enforced retirement in 1621 and his death five years later, he also published his major *History of the Reign of King Henry the Seventh* (1622); a third edition of his *Essayes or Counsells, Civill and Morall; Apophthegms New and Old*; and *Translations of Certain Psalms* (1625). Additionally, he wrote *Advertisement Touching a Holy War, Considerations Touching a War with Spain*, plus one or two other items that have since been lost.

It seems perfectly clear that Bacon was desperately trying to finish projects that had been on his mind for some years, but that he had been unable to execute due to the pressure of legal and civic responsibilities. Indeed, one of the great mysteries of Bacon's life is when he managed to do his own experimental work. There are numerous places in his writings that show Bacon had undertaken extensive experimental investigations. And certainly he writes in a familiar way about alchemical experiments and results. There are, however, no indications that he had furnaces or any kind of laboratory in his lodgings. Nor is there any indication that he worked with other alchemical adepts or performed experiments with any other investigators.

Even when his first biographer, his secretary William Rawley, tells us in *The Life of the Honourable Author* that Bacon did not get his knowledge from books, he does not go on to say that he learned by experience, but only that his knowledge came from 'some grounds and notions from within himself'. Nevertheless, we must assume that somewhere, and on many occasions, Bacon was assiduously performing experiments. As he said in the Dedication of the *Great Instauration* to King James: 'Your Majesty may perhaps accuse me of larceny, having stolen from your affairs so much time as was required for this work.'

In the *New Organon*, Bacon indicated that he thought he'd managed pretty well, all things considered:

I think that men may take some hope from my own example … If there be any that despond, let them look at me, that being of all men of my time the most busied in affairs of state, and a man of health not very strong (whereby much time is lost), and in this course altogether a pioneer following no man's track … have nevertheless by resolutely entering on the true road, and submitting my mind to things, advanced these matters, as I suppose, some little way (New Organon, *I, Aphorism 113).*

Be that as it may, within a couple of years he was

writing with renewed urgency, like a man who felt he had previously wasted most of his time in what in retrospect seemed like futile affairs of state.

It was only with a dismayed hindsight that Bacon believed he had squandered his real talents. Such was the nature of Bacon's reform programme for the sciences that it required a bureaucratic infrastructure to make it possible (see Chapter 11 for more details). Bacon's New Philosophy was to emerge as the result of a massive, well-organised collaborative effort by different specialists effectively following the principles of the division of labour. Already in 1592, while still struggling to find a patron and make something of himself, Bacon knew what he really wanted to do. He was not interested in establishing himself in the law. He wanted to be a statesman, but not one like his father. He wanted to be a statesman with a difference. Even then, as we can see from a letter he wrote in 1592 to Lord Burghley, he wanted to be a statesman of science:

I confess that I have as vast contemplative ends, as I have moderate civil ends: for I have taken all knowledge to be my province; and if I could purge it ... I hope I should bring in industrious observations, grounded conclusions, and profitable inventions and discoveries; the best state of my province.

Bacon suggested to Burghley that this was not just a matter of curiosity about nature, but a matter of philanthropy – for the benefit of all, or for the state. Although this is not sufficient to make us admire Bacon's insincere and opportunistic efforts to gain preferment, it does help us to understand why he was willing to sacrifice so much of his personal integrity for his ends.

We now know that Bacon was not exaggerating his conviction when he added in his letter to Burghley that his ambition was 'so fixed in my mind as it cannot be removed'. But we also know that nothing came of this letter to Burghley. Bacon was not set up with a unique office in the Elizabethan administration enabling him to put his ideas into action, either at this time or at any other. His diminished hopes were somewhat revived when James came to the English throne. The King had a reputation for learning and wisdom. He may have been the wisest fool in Christendom (as Henri IV of France was supposed to have said of him), but evidently wise nonetheless. Perhaps he would recognise the worth of what Bacon wanted to do? Well, he didn't; and in the end, as we have seen, Bacon was reduced to frantic efforts merely to leave sufficient materials to illustrate the kind of programme he had in mind, how it should be done and how it should work.

The envisaged massive scope of the enterprise is

nicely suggested in the title Bacon gave to his final attempt to illustrate the kind of thing he had in mind for the third part of his Great Instauration. The posthumously published example of the would-be fully comprehensive database, which was to constitute the foundation for Bacon's reform of natural philosophy, was entitled *Sylva Sylvarum*, which means 'Forest of Forests'. We all know the common expression that someone cannot see the wood, or forest, for the trees. If we wish to see the big picture, we have to recognise that the individual trees make up a forest.

In the preface to the *Great Instauration*, Bacon wrote of our need for a guide 'through the woods of experience and particulars'. His concern was to see how the trees, individual details of knowledge, made up a forest of knowledge, but he recognised that there were other forests of knowledge, composed of different trees, which combined to give an even bigger picture. There were so many, in fact, that there was a whole forest of forests. The database we require for a sound knowledge of nature must be gathered not from just one of the forests, but from the whole *Sylva Sylvarum*.

Ironically, Bacon's frantic efforts to take the Great Instauration as far as he possibly could actually led to his death. According to the often-repeated version of the story told by the English antiquary, John Aubrey (1626–97), he died as a result of experimenting with refrigeration:

As he was taking the air in a coach with Dr Witherborne (a Scotchman, physician to the King) towards Highgate, snow lay on the ground, and it came into my Lord's thoughts, why flesh might not be preserved in snow, as in salt. They resolved to try the experiment at once. They alighted out of the coach and went into a poor woman's house at the bottom of Highgate Hill, and bought a hen, and made the woman gut it, and then stuffed the body with snow, and my Lord did help to do it himself. The snow so chilled him, that he immediately fell so extremely ill, that he could not return to his lodgings (I suppose at Gray's Inn), but went to the Earl of Arundel's house at Highgate (John Aubrey, Brief Lives).

Within a few days he was dead. This almost seems too ironically pat to be true, but Bacon's most recent biographers establish that the irony really was there, even if the frozen chicken wasn't.

Bacon did die a few days after taking refuge in Arundel's house, but shortly after arriving he wrote to Arundel (who was not in residence) to apologise for imposing on the hospitality of his household. In the letter, Bacon tells us that he was out that way trying 'an experiment or two' on what he called the 'conservation and induration of bodies'. In other words, he was experimenting on the prolongation of life, always a favourite theme of Bacon's pragmatic philosophy. Bacon had recently published his *History of Life and Death* (1623), which included speculations about how life may be prolonged, and there were further considerations of the matter in the even more recently written *Sylva Sylvarum* (but this wasn't a new obsession; he'd begun writing *An Inquiry Concerning the Ways of Death, the Postponing of Old Age, and the Restoring of the Vital Powers* in 1611).

The letter suggests that Bacon had been experimenting on himself and he was taken ill with a fit of vomiting. He had good reason to experiment on prolonging his own life, since he was 65 and had been suffering from a long illness. But it sounds as though he inadvertently poisoned himself and died, either as a direct result or as a result of the extra debilitation to an already sickly frame.

It seems, then, that Bacon not only lived for the promotion of the experimental method in science, but died for it too.

• CHAPTER 5 •

'ENLARGING OF THE BOUNDS OF HUMAN EMPIRE': BACON AND MAGIC

Bacon is justly remembered for being one of the first natural philosophers to promote the idea that human knowledge and power should meet in one, and that scientific knowledge should be put to 'the improvement of man's mind and the improvement of his lot' (*Thoughts and Conclusions*, 1607). We've already noted the political and religious motivations behind this idea, but he also had a model to follow. Within the magical tradition, the knowledge of the magician was always intended to be put to use. Modern commentators have often been puzzled by Bacon's suggestion that 'Truth ... and utility are the very same things' (*New Organon*, I, Aphorism 124).

Philosophically speaking, this is a puzzling thing to say; it is easy to think of truths that are definitely not useful. However, from the point of view of the practising magician or Baconian scientist (as opposed to the speculative natural philosopher) it is easy to see that the truths they are concerned about are the ones that can be put to use. If knowledge is power, then truth is useful.

Bacon's philosophical works show that he knew a great deal of magic, that he was a practitioner of some parts of the magical tradition, particularly alchemy, and that his own philosophy was heavily indebted to magic. At the outset of Bacon's most important published work, his *New Organon* (I, Aphorism 3) he declares that: 'Towards the effecting of works, all that man can do is put together or part asunder natural bodies. The rest is done by Nature working within.'

Compare those words with these:

The whole course of Nature could teach us by the agreement and disagreement of things either so to sunder them, or else to lay them together by the mutual and fit applying of one thing to another, as thereby we do strange works.

This extract is taken from perhaps the best-known compendium of magic, the *Magia Naturalis* (*Natural Magic*, 1589), of Giambattista della Porta. Della Porta is explicitly explaining how magical effects take place – bodies have occult powers by which they can act upon one another, sometimes reinforcing one another (when their powers are in agreement), and sometimes cancelling one another out (if their powers are in disagreement).

The same idea can be seen in another leading writer in the magical tradition, Cornelius Agrippa:

Magicians are careful explorers of nature, only directing what nature has formerly prepared, uniting actives to passives and often succeeding in anticipating results so that these things are popularly held to be miracles when they are really no more than anticipations of natural operations (Of the Uncertainty and Vanity of the Sciences, *1526).*

Seemingly miraculous results can be accomplished, Agrippa says, simply by combining or uniting the right substances: actives to passives. Similarly, strange works can be performed, according to Della Porta, simply by applying one thing to another. When Francis Bacon said that to bring about desired effects all that man could do was to 'put together or part asunder natural bodies', and then rely on the powers of Nature to accomplish it, he was writing like a magician.

Underlying this view was the magical belief in sympathies and antipathies – the belief that bodies, inanimate as well as animate, were favourably disposed to some other bodies or kinds of bodies and unfavourably disposed to others. This is what Della Porta means by the agreement and disagreement of things. By bringing sympathetic bodies together, a particular effect could be enhanced. Alternatively, if the power of something needed to be weakened or nullified, this could be done by introducing something known to be antipathetic towards it. As Bacon wrote in his *History of Density and Rarity* (1624):

> *Just as bodies lay themselves open all round to attractive and friendly things and go to meet them, so when they happen on things hateful and hostile they fly from them all round and pull back and withdraw into themselves.*

Similarly, in his *New Abecedarium of Nature*, Bacon insisted that: 'We must … investigate the individual and particular friendships and quarrels or sympathies and antipathies of bodies with diligence and care, seeing that they bring with them such a number of useful things …' Bacon even planned to write a *History of Sympathetic and Antipathetic Natures*, but failed to get further than the preface.

To us, the use of terms like 'sympathy', 'friendly', 'hateful' and so on seems highly incongruous when discussing inanimate objects, but the underlying meaning is not so different from what we believe today. To say that iron has a sympathetic attraction to the lodestone or magnet compares quite well with the modern claim that iron is attracted to the magnet because it is paramagnetic. Furthermore, it is easier to justify. If we wish to get beyond the merely descriptive defining term (let's face it, to say something is paramagnetic is merely to say it is attracted to magnets) and try to understand why a paramagnetic nature makes a body move in a magnetic field, we will find ourselves on a lengthy and difficult quest with no guarantee of real intellectual satisfaction at the end. This does not mean that modern science is misconceived or fraudulent, but it does indicate that we take a great deal of it entirely on trust.

We believe things about magnets and magnetism, and much else besides, without really understanding how they work. This is true not just of laymen but of some scientists too, even in many cases scientists who might use the theory of magnetism or some other improperly understood phenomena in their own work. The theory behind the belief in sympathy and antipathy was entirely straightforward, however.

It was always assumed that God had created the world according to a hierarchical scheme or plan. In the Great Chain of Being (as it was known), the most inferior creatures dwelled near the bottom; gradually coming up the scale, creatures were increasingly complex or advanced and increasingly admirable or noble. Down at the bottom were slugs and worms and other vile things, while up at the top, in the earthly part of the scale anyway, was man (woman was always below man in those days). Beyond man, and the earthly realm, creatures continued to advance higher on the scale. There were nine orders of angels, from the common or garden angel at the bottom, through archangels, principalities, powers and on to cherubim and seraphim at the top, immediately below God Himself.

But it was recognised that all the creatures of the world could not be placed in a simple hierarchical scheme. There had to be complexities within the chain to reflect the nature of things. There were, after all, three Kingdoms, representing the three broad kinds of God's creatures: animal, vegetable and mineral. Gold was the noblest of the metals in the mineral kingdom, but was it nevertheless inferior to the lowliest form of plant life or inferior to the most repugnant slug? Surely not.

On the contrary, the Great Chain of Being was full of what were known as correspondences. Gold, the noblest metal, corresponded to the vine, noblest among plants, and to the lion, king of the beasts. But it also corresponded to the Sun, the golden orb in the heavens and the noblest of the heavenly bodies. Likewise, it also corresponded to kings and princes, noblest among men. Silver corresponded to the Moon, to wheat, and the eagle, second only to the lion in regality among beasts, and to queens (see Figure 3).

*

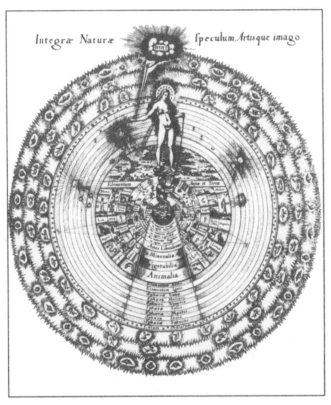

Figure 3: This complex image is meant to depict the Great Chain of Being from the hand of God down to Mother Nature. Additionally it shows a chain connecting Nature to man, the 'ape of nature', so-called because he imitates natural processes through his arts. The central, concentric images indicate some of the supposed correspondences between creatures in the Great Chain of Being – man, vine, the Sun, gold; woman, wheat, the Moon, silver. The illustration is from the first volume of Robert Fludd's *Utriusque Cosmi Historia* (*History of Both Cosmoses*, Oppenheim, 1617).

This way of thinking about the world clearly served a cultural and political function. These were times when the rich man was in his castle, the poor man was at the gate and everyone knew their place. But it also served a magical function. The various correspondences between things in the Great Chain of Being, whether animal, vegetable or mineral, accounted for the supposed sympathies between things, and by contrast for the antipathies between things that corresponded to opposed creatures in the hierarchy.

It was the multiplex of correspondences between different links in the Great Chain of Being that underwrote sympathetic magic. Given the fact that the magician had no powers of his own, but could only exploit the sympathies and antipathies between things and bring together or separate interacting bodies or substances, it was important for the magician to know the correspondences. This is what the English poet John Donne was referring to in his mournful poem 'Anatomie of the World' (1611):

What Artist now dares boast that he can bring
Heaven hither or constellate any thing,
So as the influence of those starres may bee
Imprison'd in an Hearbe, or Charme, or Tree,
And doe by touch, all which those stars could do?
The art is lost and correspondence too.

If your stars were not favourable, an artist or magician should be able to overcome this by using herbs or charms that corresponded to the benign heavenly bodies, absent from your horoscope, applying them in some way to bring about the same benign effect. But Donne is pessimistic. He believes the art of magic is lost, because knowledge of the correspondences is lost. You need knowledge to have power.

Francis Bacon was evidently much more sanguine. Even if the correspondences were lost, they could be rediscovered. In his *New Organon* (II, Aphorism 31) he advocates a search through magical lore for knowledge of useful influences:

> *For although such things lie buried deep beneath a mass of falsehood and fable, yet they should be looked into ... for it may be that in some of them some natural operation lies at the bottom; as in fascination, strengthening of the imagination, sympathy of things at a distance, transmission of impressions from spirit to spirit no less than from body to body, and the like.*

At the very end of the *New Organon*, Bacon emphasises magic as one of the most important means of discovering 'the virtues and actions of bodies'. Of those different kinds of phenomena of nature that he has discussed in the book, he singles out just a few as more important than the rest. Among these is magic, from the study of which 'there cannot but follow an improvement in man's estate, and an enlargement of his power over nature'.

Magic was always regarded as an art rather than a science. Its business was not the understanding of nature for its own sake, but to put nature and its powers to use for the benefit of the magician or his patron or client. It is hardly surprising, therefore, that Bacon, with his concern for the improvement of the human condition, should be drawn to magic.

Although Bacon has been described as a 'philosopher of industrial science',[1] and presented as a thinker who took inspiration from contemporary craftsmen and artisans and their technical innovations, it is very obvious that he is far less interested in technological change than in more magical visions. Consider, for example, his list of desiderata, entitled *Great Works of Nature for the Particular Use of Mankind* (*Magnalia Naturae Praecipue Quoad Usus Humanos*, 1624). Virtually all the entries are straight out of the catalogue of things that magicians had always claimed they could do: the prolongation of life, the increasing of strength and activity, the altering of features, conversions of bodies into other bodies, force of the imagination (either upon another body or upon the body of the imaginer), impressions of the air and raising of tempests, deceptions of the senses – and so the list goes on.

Bacon did not write like an engineer or a technologist, but like a magician.

In fact, there was no such thing as industrial science or engineering in Bacon's day. Bacon frequently talks of mechanics, but at the beginning of the seventeenth century even this was a very new domain. The study of machinery, which depended on moving parts as well as the principles of statics and the understanding of movement and force, was still only an emergent discipline at this time. What's more, like alchemy or astrology, it was separating itself out of the natural magic tradition.

Throughout the Middle Ages the design and use of machinery was regarded as part of the province of the natural magician. The reasoning behind this was no more profound than that, because machines worked by hidden and not obvious means, they were occult objects. Machinery was always associated with magic. One of the most famous sources for such an association was found in a major magical text by one of the most notorious magicians of the Middle Ages, Roger Bacon (*c.* 1214–92), a Franciscan monk and one of the supposed models for Dr Faustus (a legendary magician who sold his soul to the devil in exchange for knowledge and power), who was no relation to Francis.

Instruments of navigation may be made without men for rowing, so that great ships, both river and sea going, may be borne along with just one man steering, and with greater speed than if they were full of [rowing] men. Also chariots can be made to move without an animal with inestimable force, as we reckon the scythed chariots to have been which were used for fighting in ancient times. Also there can be made instruments of flying with a man sitting in the middle of the instrument and revolving a contrivance by means of which artificially composed wings beat the air, like a flying bird ... Instruments may also be made for walking in the sea or in rivers, right to the bottom, without any bodily danger. For Alexander the Great used these for seeing the secrets of the sea according to the account of Ethicus the astronomer. These were made in ancient times, and have been made in our times as is certain, unless it be the instrument for flying, which I have not seen, nor do I know anyone who has seen it, but I know a wise man who has thought out how to bring about this artifice (Communia Mathematica, *c. 1272).*

Needless to say, in spite of Roger Bacon's claims, none of these things really existed, but it was the kind of thing that people believed magicians could do. Although Francis Bacon was aware of the new discipline of mechanics, he, like many of his contemporaries, still tended to see marvellous machines as the province of the magician. In his influential utopian fiction *The New Atlantis* (1627), for example, the riches of Salomon's House, Bacon's vision of an ideal scientific research establishment, included similar ideas. 'We imitate also flights of birds', the Master of Salomon's House declares:

[W]e have some degrees of flying in the air; we have ships and boats for going under water, and brooking of seas; also swimming girdles and supporters. We have divers curious clocks, and other like motions of return, and some perpetual motions.

In context, this part of Francis Bacon's story is reminiscent of earlier magical writings like Roger Bacon's, rather than like some up-to-the-minute technical account. This impression is reinforced by the fact that shortly afterwards Bacon talks about the oath of secrecy sworn by the fellows of Salomon's House, and their deliberations about whether or not to publish details of their inventions. This kind of secrecy would certainly have been seen by Bacon's contemporaries as typical of magical adepts.

The whole point of magic, then, was to exploit natural phenomena, the natural powers of bodies, their forces and movements, in order to bring about practical ends. It is hardly surprising, therefore, that Bacon should turn to it as a major source of inspiration for his own attempts to reform the ethos of natural philosophy. But the influence of magic on Bacon did not end there. As soon as Bacon turned to magic he could hardly fail to notice that its principal method was experimental.

• CHAPTER 6 •

'EFFECTING OF ALL THINGS POSSIBLE': MAGIC AND THE EXPERIMENTAL METHOD

As Bacon saw it, 'The aim of magic is to recall natural philosophy from the vanity of speculations to the importance of experiments' (*The Proficiency and Advancement of Learning*, Book 1, 1623). Natural magic was what Bacon called an operative part of natural philosophy, as opposed to a speculative part, and although some of its achievements had been arrived at by chance, others were discovered by 'a purposed experiment', or 'an intentional experiment' (*Advancement of Learning*, 1605).

Bacon was highly critical of natural magic, seeing its achievements as disappointing to say the least, but he was in no doubt that its method, if properly applied, was superior to the methods of speculative philosophy. The way to knowledge indicated by speculative philosophy was erroneous and impassable, he insisted, being based only on the 'whirl and eddy of argument' and the 'mist of tradition' (*New Organon*, I, Aphorism 82).

Consisting of nothing more than personal meditation on everything that has been said on a particular topic before, the results of this method are mere opinions based on insecure foundations. The result of this can only be the spinning of cobwebs of learning out of one's own mind, with little or no basis in reality. We should not wonder that the true goal of the sciences has not yet been achieved, because philosophers have gone astray by abandoning experience.

Like others among his contemporaries, Bacon saw this as the result of a belief originating with the Ancient Greeks that:

> [T]he dignity of the human mind is impaired by long and close intercourse with experiments and particulars, subject to sense and bound in matter; especially as they are laborious to search, ignoble to meditate, harsh to deliver, illiberal to practise, infinite in number, and minute in subtlety (New Organon, I, Aphorism 83).

The upshot is that 'the true way', of experiment and experience, has not merely been abandoned but 'rejected with disdain'. However, if this was true of the natural philosophy tradition, particularly as it was conducted by scholastic philosophers within the universities, it was not true of the completely separate tradition (disallowed within the universities) of natural magic. 'Magic', Bacon wrote, 'proposes to recall natural philosophy from a miscellany of speculations to a magnitude of works' (*Advancement of Learning*).

Since magic was chiefly concerned with exploiting the sympathies and antipathies between corresponding things in the Great Chain of Being, and since the assumption was that these powers of agreement and disagreement were hidden or occult, the magician could only discover the powers of things empirically. Knowledge of what actives to combine with what passives, as Agrippa put it, could only be acquired by trial and experience. A leading Italian Renaissance magus (magician), Pietro Pomponazzi, summed up what was required:

> *If the correspondences are real, it follows also ...*
> *that there are herbs, stones, or other means of this*
> *sort which repel hail, rain, and winds; and that*
> *one is able to find others which have naturally*
> *the property of attracting them* (On the Causes
> of Natural Effects, *1520).*

So, if there are correspondences, we should be able to find herbs, stones and other natural objects that are antipathetic or sympathetic to hail, rain, winds (or whatever). Then we should be able to induce hail and rain or drive them away. But, as Pomponazzi pointed out, this assumes that 'men are able to discover [the sympathies and antipathies] naturally'. If so, then 'magick', as Giambattista della

Porta wrote, 'is nothing else but the survey of the whole course of Nature'.

> *For, whilst we consider the heavens, the stars, the elements, and how they are changed, by this means we find out the hidden secrecies of living creatures, of plants, of metals, and of their generation and corruption; so that this whole science seems merely to depend upon the scrutiny of Nature* (Natural Magic, *1589*).

Or, as Agrippa said, '[N]atural magic is that which, having contemplated the virtues of all natural and celestial things and carefully studied their order, proceeds to make known the hidden and secret powers of nature' (*Of the Uncertainty and Vanity of the Sciences*). Small wonder that successful magicians have to be, as he went on, 'careful explorers of nature, only directing what nature has formerly prepared'.

Needless to say, successful magicians were thin on the ground. The kind of empirical knowledge demanded by the theory of magic (even if the theory had been true) was almost impossible to achieve. For the most part, magicians relied on magical compendia, textbooks of magic, full of traditional but untested lore. This was a far cry from the empirically learned magus of legend. Those dissatisfied with this kind of book learning could call upon another aspect of magical theory to provide short-cuts to the right kind of empirical knowledge, but in reality this must have proved misleading as often as it was helpful. This was the belief in signatures.

It was assumed that when God created the world, establishing the correspondences between the different scales of nature within the Great Chain of Being, He left signs or signatures on things to indicate the correspondences. The walnut provides one of the best examples. Crack open the shell of the walnut and the flesh of the nut appears to be divided down the middle and to have a surface made up of convolutions (you can try this experiment at home). Doesn't it look very like the human brain inside the skull? The walnut has a signature that makes it look like the human head and brain.

It was a general assumption in those days, before the onset of secularisation, that God would not do anything in vain. There was always a reason for everything He did. It was not always possible to fathom God's mind, of course, but the signatures seemed to be obvious signs of correspondences. The walnut must correspond to the human head. All that remained was to empirically test to see whether it could be used as a cure for, say, headaches, or insanity, or whatever, and how it should be administered (as a poultice, an ointment, or to be taken internally in one form or another).

As you can imagine, this was hardly likely to meet with much success. It's likely that walnuts have no special beneficial effects on ailments of the head or brain anyway, but it's perfectly likely that a headache will cease after eating walnuts and some cases of mental illness might even disappear after dosing with walnuts. Opinions are likely to be divided as to the efficacy of walnuts, therefore, and the probable outcome will be confusion.

But human ingenuity and the potency of experience should not be unduly dismissed. After all, there were many folk remedies that are now known to work and that, in some cases, have been incorporated into modern medicine. Salicylic acid, the active ingredient in aspirin, is now known to occur in the sap of the weeping willow. Village cunning men and wise women or witches did not know about salicylic acid, but they did know how to extract a pain-killer from weeping willow (its look of bowed weeping being seen as a sign from God of its helpfulness in this regard), using it as a cure for headache, toothache and other pains.

Similarly, the anti-depressant qualities of the herb St John's Wort have now vindicated its traditional use as a cure for melancholy. The list could easily be extended. There can be little doubt that, in spite of the inherent difficulties, the empirical approach of the magical tradition proved more beneficial to mankind, as Bacon recognised, than the official learning of the universities and medical schools, where book-learning was all that seemed to count.

Joan Baptista van Helmont, a leading Flemish alchemist and healer in the magical tradition, tells us in a brief autobiographical account that botany in the universities was taught as though it had in no way progressed since the days of Dioscorides (*fl.* first century AD) and his encyclopaedia of *materia medica* (medical substances). Indeed, Van Helmont wrote:

> *Even today, his pictures, nomenclature and descriptions of plants are after all these years the recognised basis of every discussion, but ... no advance had been added, about their virtues, properties and use ... But I knew that about two hundred plants, though identical in quality and grade, are quite different in virtues, and that a number of others different in quality and grades act synergically. So not the herbs (the signs of Divine love), but the herbalists fell into disrepute with me* (The Origin of Medicine, *1648).*

It was possible, therefore, for the experienced magus to glean useful knowledge for the benefit of mankind, in spite of the problems inherent in an empiricism guided by nothing more than a notion of divine signatures. Bacon must surely have recognised this. For one thing, he subscribed to a philosophical system based closely on the ideas of Paracelsus (another alchemically and magically inspired healer, like Van Helmont), whose experience in the field proved highly successful. Nevertheless, Bacon was concerned to establish himself as offering something new, and so he did not hesitate to point out the inadequacies of the empiricism of the alchemists and the natural magicians.

> For the Alchemist nurses eternal hope, and when the thing fails, lays the blame upon some error of his own; fearing that he has not sufficiently understood ... or else that in his manipulations he has made some slip of a single scruple in weight or a moment in time (whereupon he repeats his trials to infinity) ... Again, the students of natural magic, who explain everything by Sympathies and Antipathies, have in their idle and most slothful conjectures ascribed to substances wonderful virtues and operations; and if ever they have produced works, they have been such as aim rather at admiration and novelty than at utility and fruitfulness (New Organon, I, Aphorism 85).

This kind of criticism of magic was more often than not entirely justified. Not all alchemists were as accomplished and assiduous as Paracelsus or Van

Helmont. Indeed, many, as everyone knew, were nothing more than frauds and charlatans, whose ambitions were set no higher than to gull a gullible public. Others were deceived by their own pride, 'conferring the title sacred upon certain fleeting meditations instead of reserving it for the divine signature on things', Bacon wrote in his *Refutation of Philosophies*. Although Bacon was always ready to point to its faults, often in the most vigorous and contemptuous terms, there can be little doubt that he nonetheless recognised the real value of the magical tradition. As he willingly accepted, astrology, natural magic and alchemy were noble and worthwhile pursuits even though, in practice, they were full of error and futility.

Apart from its professed aim of using knowledge of nature for practical benefit, there was another reason for Bacon's admiration of magic. He lived through the last stages of the period of history known as the Renaissance, and was steeped in Renaissance values. The Renaissance was given its name by the intellectuals of the day who saw themselves as living in a period of re-birth. The re-birth they had in mind was the re-birth or renaissance of the glories of Ancient wisdom. A few chance discoveries of Ancient writings, which had been preserved in monastery libraries, initiated a systematic search of monasteries all over Western and Eastern Europe (even in the days long before the Iron Curtain,

Western Europe and Eastern Europe were divided by religion – Roman Catholic or Eastern Orthodox – and by the *lingua franca* among administrators and intellectuals – Latin in the West, Greek in the East). The result was an astonishing recovery of Ancient works that had long been assumed lost, or the existence of which had never been dreamed of.

The significance of these literary discoveries was enormous. Virtually all the Ancient writings available to us today were discovered by Renaissance scholars. A few things were known before, by a very restricted handful of Ancient writers, and one or two items have been recovered since the Renaissance, but most were found in the fifteenth and sixteenth centuries. However, Renaissance scholars were not simply excited by the fact that they had found old texts. They were excited because they believed that this Ancient wisdom represented a major step towards the recovery of the wisdom of Adam. It had long been assumed in the Judaeo-Christian tradition that Adam had known all things, but that this knowledge had gradually been forgotten after the Fall, when Adam and Eve had disobeyed God by eating the forbidden fruit and had been cast out of Eden. Ancient wisdom was admired not just for antiquarian reasons, therefore, but because it was regarded as wisdom that was preserved before too much of the Adamic wisdom, directly God-given, had been forgotten.

We know that Bacon subscribed to the belief in Adamic wisdom, and the magical belief that Adam could command things because he named them (see Genesis, 2, 19). As he wrote in *Valerius Terminus* (1603), '[W]hensoever he [meaning man] shall be able to call the creatures by their true names he shall again command them.' The true names, invested with a power of invocation over the creatures, were those first given to the creatures by Adam. Clearly, it would be of immense usefulness if the Adamic names could be recovered.

It just so happened that there was a set of newly discovered writings that were regarded as particularly close to the Adamic wisdom. These were the writings of the Ancient Greek god Hermes, known as Trismegistus, thrice-great (since he was a great ruler, a great religious teacher and a great magus). We now know that the writings that make up the so-called *Hermetic Corpus* were written in the first and second centuries after Christ by Neoplatonic writers, some of whom were influenced by or sympathetic to Christian beliefs. In order to make their writings seem more important than they were, however, the authors of these magical texts claimed they were written by Hermes Trismegistus. The Renaissance scholars who rediscovered these writings fell for it hook, line and sinker. Assuming that these writings, which obviously included Christian ideas, were written long before the Christian era

– since pre-Christian Greeks worshipped Hermes as a god – Renaissance intellectuals believed they had found a pagan strand of Adamic wisdom as old as the Pentateuch of Moses, which represented the oldest Judaeo-Christian strand of Adamic wisdom.

It was in this way, therefore, that a group of magical texts became revered among Renaissance intellectuals as the oldest teachings (and therefore the closest to the God-given wisdom of Adam) available outside the Holy Scriptures. After centuries in which the Church had been denouncing magic as nothing more than a set of tools for deception used by the Devil, magic suddenly came to be seen as a supreme form of wisdom. It now became possible to denounce the Church for its jealous repudiation of magical power. As Cornelius Agrippa wrote:

The outstanding question is this: why is it that although magic originally occupied the pinnacle of excellence in the judgement of all the ancient philosophers and was always held in the highest veneration by those great sages and priests of antiquity, subsequently (from the beginning of the rise of the Catholic Church) it became an object of hatred and suspicion to the holy Fathers, and was at last hissed off the stage by the theologians, condemned by the sacred canons and, in the end, outlawed by the judgement of all laws? (Dedicatory Letter to Johannes Trithemius, *1533*)

At the end of the Renaissance, Bacon was therefore well placed to recognise the value in the magical tradition and to be inspired by its emphasis on using knowledge to gain mastery over nature, and its emphasis on experience and experiment to acquire that knowledge.

But if the experimental method was already fully accepted and in use in the magical tradition, why do historians make such a fuss about the fact that Bacon promoted the experimental method? To understand this we have to be aware of the rigid separation, in Bacon's day, between natural philosophy and natural magic. What was innovative about Bacon was that he advocated the use of the experimental method of the magician in a reformed natural philosophy. Where natural philosophy and natural magic had once been completely separate traditions – the one rationalist and speculative, the other experimental and pragmatic – Francis Bacon suggested bringing them together.

Although the phrase 'natural philosophy' is often used by historians as though it is just an old-fashioned term for what we call science, and is interchangeable with it, this is in fact highly misleading. Natural philosophy, before Francis Bacon and other seventeenth-century thinkers changed its character beyond all recognition, was not like modern science. It became closer to modern science, thanks to Bacon and others, when it was amalgamated with precepts and practices taken from the magical tradition. The most important of these were the precept that knowledge should be practically useful, and the practice of gathering knowledge by experiential and experimental means.

Bacon was by no means the only thinker who promoted magic over natural philosophy. But he *was* the only one who did it within the context of an explicit programme for reforming natural philosophy. And this is where his importance lies. Other thinkers inspired by the magical tradition used their experimental knowledge to help them establish a new system of belief about the natural world, a new cosmology (in the broadest sense, meaning a theory of life, the universe and everything, not just a theory of the organisation of the heavenly bodies).

Paracelsus and Joan Baptista van Helmont both developed alchemical cosmologies, in which all things were seen in terms of alchemical processes – from the creation of the solar system out of an original chaos, to the digestion of one's dinner, from the working of the imagination to the formation of stones and other minerals. William Gilbert, a contemporary of Bacon's who used a whole battery of experimental techniques to investigate the nature of magnets and prove that the Earth itself was a giant magnet, spent no time at all commenting on the fruitfulness of his experimental method, but went straight on to develop a whole cosmology based on a magnetic philosophy. Needless to say, this rapidly became highly speculative, unlike his work on magnets.

In many respects Bacon was no different from these other thinkers. He too had his own system of cosmology, based chiefly on alchemical and magical ideas – particularly those of Paracelsus (indeed, it has been called, with good reason, a 'semi-Paracelsian cosmology'). Bacon was indefatigably committed to this magical cosmology from 1612 at the latest (although elements of it can be found as early as 1592) and remained so until his death. Many of his own works were written to provide evidence for and confirmation of this cosmological theory.

In view of this long-standing commitment it seems safe to assume that Bacon, no less than Paracelsus or Gilbert, believed he had found the correct and true understanding of nature and its processes. He might, therefore, have spent his time trying to persuade his contemporaries of the truth of his new system of philosophy. Had he done so, he would no doubt have been remembered in the history of science as just another system-builder of the late sixteenth and early seventeenth centuries, whose ideas were interesting but misconceived.

But he is not remembered this way. He is remembered for the very reason that set him apart from the other magically inspired writers of the late Renaissance. His great achievement, so characteristically English, was to hold it always in his mind that he might just be mistaken. Instead of promoting his own system of belief, therefore, he directed most of his energies to promoting a new pragmatic purpose for the study of nature, and a new method of doing science.

By introducing pragmatism and experimentalism from magic into natural philosophy, Bacon radically changed natural philosophy and made it potentially far more fruitful. Instead of trying to persuade his contemporaries that he had discovered the true system of cosmology, he tried to outline for them the best procedures for understanding the operations of the natural world. In short, he tried to show them the importance, in its own right, of developing a scientific method. In this way, then, it could be said that Bacon really did succeed ...

in kindling a light in nature – a light which should in its very rising touch and illuminate all the border-regions that confine upon the circle of our present knowledge; and so, spreading further and further should presently disclose and bring into sight all that is most hidden and secret in the world ('Proemium', Of the Interpretation of Nature).

• CHAPTER 7 •

'SEEKING FOR EXPERIMENTS OF LIGHT': BACON AND THE DECLINE OF MAGIC

In spite of the obviousness of Bacon's use of magical ideas, whether in his theories on method or in his own semi-Paracelsian cosmology, there has been a tendency among historians of science to play this down. The reasons for this are easily stated. Magic is seen on the one hand as being a thoroughly irrational pursuit and on the other as being based on assumptions about supernatural powers. The idea that anything so irrational could have influenced something as supremely rational as modern science seems immediately untenable. Similarly, how can a system of belief that relies upon supernatural powers have had anything to do with an understanding of the natural world? The whole point of supernatural powers is that they are brought in to explain or achieve whatever cannot be understood or accomplished naturally. Supernaturalism, therefore, is the enemy of real scientific understanding.

If Bacon is to be retained as a major figure in the history of science, then the role of magic in his thinking has to be minimised. Fortunately for historians writing these kinds of accounts, Bacon was more often than not a vicious and contemptuous critic of magic. This makes it easy to say that although there are some indications that he was initially inspired by magic, he rejected it as puerile and misguided, and went on to develop his own original methodology, which has proved to be one of the great foundational cornerstones of modern science.

It should be noticed, however, that Bacon never dismisses magic on the grounds that it is irrational, nor on the grounds that it relies on the supernatural. What bothers modern historians about magic clearly did not bother Bacon. For Bacon there were two main problems. First, that magic was all too easily exploited by mountebanks and swindlers, thus compromising whatever useful information can be gleaned from it. Second, that its experimental method was not sufficiently systematically pursued by its practitioners, and that its results were too often noted simply for their usefulness, without any attempt to understand the natural processes by which these results were brought about. Magicians, as Bacon complained, sought only for 'experiments of fruit', while he was 'seeking for experiments of light' (*New Organon*, I, Aphorism 121).

It is not hard to understand why there is this discrepancy between what modern historians think is wrong with magic and what Bacon thinks is wrong with it. The simple truth is that modern historians have not properly understood the magical tradition. In particular they have failed to grasp what was meant by the term 'natural magic'. There was nothing irrational in Bacon's day in believing in magic. The belief that bodies had specific properties and virtues that enabled them to act upon or interact with other bodies to bring about particular ends entirely conforms with modern beliefs.

While today we would not think of such activities in terms of sympathies and antipathies (although, by the way, chemists do still talk about chemicals having different levels of affinity for one another), this does not mean this belief was irrational. You can call it wrong, if you like, but not irrational. It was, after all, based on a belief in corresponding planes along the Great Chain of Being, so there was a logic behind any claims about sympathies.

There was nothing irrational about the belief in the Great Chain of Being, either. On the contrary, as King Lear was made by Shakespeare to discover, to deny what then seemed to be the natural hierarchies in things (in Lear's case, the father's place as the head of the household and the King's place as the head of the kingdom) was to descend beyond the merely irrational to sheer madness. Nor was it irrational to believe that by intoning a special combination of words, a magic spell or incantation, remarkable physical effects could be produced.

This was a time, don't forget, when everyone believed in God and religion. Nobody doubted that God could create light simply by saying, 'Let there be light', and most people (namely, all Roman Catholics) still believed that the priest could turn wine into Christ's blood and lots of wafers of bread into Christ's flesh, simply by saying a few words over the elements of the Eucharist. You can call this an irrational belief if you like, but I subscribe to the contrary view that, since you could be burned at the stake for denying the priests' powers, you'd be mad not to go along with the belief. And, if you do believe this, it's not unreasonable to speculate that the priests' powers may be emulated by magic spells. God might have invested words with a natural power of their own, just as He invested weeping willows with a natural power of easing pain, and magnets with a power of attracting iron. Finally, it was not irrational to believe in demons and angels at a time when these were important figures in orthodox religious teaching.

Neither Bacon nor his contemporaries would have thought of magic as irrational. They did not think of it as supernatural either. The pre-eminent form of magic was always natural magic, so called because it was based on assumptions about the natural properties and powers of bodies. Magic, according to Martin del Rio in his *Disquisitions on Magic* (1608), is 'an art or technique which by using the power in creation rather than a supernatural power produces various things of a marvellous and unusual kind, the reason for which escapes the senses and ordinary comprehension'.

What we all too often imagine when we think of magic is the wizard standing inside a pentangle, with his pointed cap and a wand, summoning up a demon. We usually assume, in such imaginative scenarios, that the demon will be invested with supernatural powers and that by merely extending a hand he can shoot bolts of force out of his fingers that can simply kill, or perhaps completely transmute something, or perhaps turn something invisible – you name it, these bolts of force can do it at the demon's whim. If you were able to present such a scenario to a contemporary of Bacon's, he or she would be utterly perplexed as to how you came up with such unbelievably crazy ideas. He or she would also be very worried about the prospects for your eternal soul.

For pre-modern thinkers, only God could perform supernatural acts. He could, of course, supernaturally arrange for priests, preparing for the mass, to be able to perform supernatural acts too, but nobody or nothing could perform supernatural acts by their own power. To imagine that they could was to deny God's superiority and was, therefore, a dreadful blasphemy. It was especially true that demons, including the Devil himself, had no supernatural powers. They, like you and me, were creatures of God, and as created beings they were part of the natural world. What's more, although God might temporarily allow an Angel some supernatural power, He would never allow this for a fallen angel, a demon.

So, although today we tend to think of demonology as the pre-eminent form of magic, this is an anachronistic misconception. For pre-modern thinkers, the summoning of demons was simply a highly dangerous short-cut to discovering the truths of natural magic. Remember how difficult it is to discover by trial and error the sympathies and antipathies of things – even in cases where there seem to be clear indications provided by the signatures of things (such as the similarity between the walnut and the brain in the skull). One way around this difficulty was to summon a demon and ask it to accomplish what you want, or to furnish you with the relevant information. By summoning a demon you could have the benefit of the demon's expertise in natural magic, but that's all. The Devil himself had no power of his own, merely a profound knowledge of how to accomplish things using natural magic. As William Perkins explained in his *Discourse of the Damned Art of Witch-Craft* (1608):

> *[The Devil has] exquisite knowledge of all natural things, as of the influences of the starres, the*

constitutions of men and other creatures, the kinds,
vertues, and operations of plantes, rootes, hearbes,
stones etc. which knowledge of his goeth many
degrees beyond the skill of all men, yea even those
that are most excellent in this kind, as Philosophers
and Physicians are.

You could say the Devil was the first scientist. (Perkins left out magicians from the company of philosophers and physicians, by the way, because, as a clergyman, he didn't want to acknowledge any excellence in magicians.) Similarly, John Cotta, in *The Triall of Witch-Craft* (1616) insisted that:

Though the Divell indeed, as a Spirit, may do and
doth many things above and beyond the course
of some particular natures: yet doth he not, nor
is he able to rule or command over generall Nature,
or infringe or alter her inviolable decrees in the
perpetual and never-interrupted order of all genera-
tions ... For Nature is nothing else but the ordinary
power of God in all things created, among which
the Divell being a creature, is contained, and there-
fore subject to that universall power.

So, because he is a spiritual being, the Devil could no doubt fly or pass through walls, or do similar things that are above and beyond what non-spiritual things can do, but there was much else besides that he could *not* do. He couldn't, for example, make a broom fly or a witch pass through a wall.

The fact is, then, that current common assumptions about the nature of magic are a long way from the truth of things in the sixteenth and seventeenth centuries. Our conception of the world of magic is highly coloured by fantasists from the eighteenth and especially the nineteenth centuries (to say nothing of more recent fictions taking their starting point from the nineteenth-century fantasies). The seemingly all-powerful demons of modern myth are, ironically, the product of secular imaginations.

The imaginations of pre-modern and early modern thinkers were reined and disciplined by their religious beliefs. Such religious believers would never allow supernatural powers to demons. Everything in nature was created by God and by definition only God was above nature. Furthermore, because the power of demons was believed to derive only from their superior knowledge of *natural* magic – their knowledge of the *natural* powers of things – demonology was always a subsidiary aspect of magic, subsidiary to natural magic.

By summoning a demon a magician could not accomplish anything that another magician might have been able to accomplish by natural means, since all the demon could do was exploit natural magic. But the magician who summons a demon is also endangering his immortal soul. It is one thing to exploit a demon, but if the demon contrives to exploit you then you are all too likely to end up among the damned. This was the main reason for the Church's opposition to magic – it was seen as a damned art. The more successful a magician was, the more likely it seemed that he had sold his soul to the Devil.

It is no accident that our modern, erroneous, view of magic differs so much from the natural magic of the pre-modern era. During the seventeenth century, major aspects of the natural magic tradition were amalgamated with natural philosophy to provide a new philosophy of nature that combined the pragmatic utility and experimentalism of magic with the rationalism and concern to understand causes of the natural philosophical tradition. The result was something much more recognisably like modern science.

Although by no means the only thinker involved in this process of amalgamation, Francis Bacon was undoubtedly the leading figure. Where others in this movement towards amalgamation taught by example, showing how certain magical ideas could be useful in understanding specific aspects of the world (thinkers such as Paracelsus, Van Helmont, Gilbert, Johannes Kepler, William Harvey, Robert Boyle, Robert Hooke and Isaac Newton), Bacon chose instead to talk more generally about the ethos and method of natural magic and its usefulness for reforming what many saw as the moribund and misconceived traditional natural philosophy.

But the amalgamation of natural magic with natural philosophy was not an event, it was a long and slow process. During the process, many aspects of natural magic were rejected and discarded. Alchemy, a major branch of natural magic, became closer to our idea of chemistry as its more symbolic and emblematic aspects were discarded (because of the belief in correspondences, the process of turning lead into gold, for example, was traditionally seen as a process bound up with the ennoblement of the alchemist, but this idea was jettisoned).

Astrology, another major aspect of the magical tradition, looked for a while as though it might make it across into the new science, but by the beginning of the eighteenth century it was effectively excluded. Many, but by no means all, of the ideas inherent in sympathetic magic were absorbed into science. The issue here was often one of experimental confirmation, or otherwise of putative natural effects. But often it was hard to decide, and many physical effects remained controversial.

Consider the example of the weapon salve. This was a famous, or notorious, magical ointment developed and promoted by the Paracelsians (followers of Paracelsus). It was an ointment that could cure wounds incurred in battle by applying the ointment to the weapon that caused the wound. That's right – the ointment was not put on the wound but on the weapon. If, as seems likely, the weapon was not available, then the ointment was applied to a bandage or piece of cloth that had been soaked in blood from the wound but which was then kept away from the injury. This proved to be much more successful than the standard means of treating wounds. No, really, it did. To understand this you just need to know what the standard method of treatment was.

Since time immemorial, healers had noticed that wounds run with yellow matter, pus, before they heal. We are speaking, of course, of the days before antiseptics – indeed before anybody ever dreamed there were such things as invisible germs and before any concomitant ideas of the need for hygiene in this regard. Formation of pus in a wound seemed, therefore, to be part of the healing process. Here, then, we have a perfectly reasonable conclusion based on often-repeated observational evidence. This was the medical theory known, believe it or not, as the theory of 'laudable pus': pus is a good thing, a sign that healing is taking place. Accordingly, doctors and surgeons throughout the ages would pack wounds with irritating substances – ground up egg-shells, sand, coarse feathers and the like – in order deliberately to stimulate the formation of pus and thereby, as they thought, *accelerate* the healing process. In the case of serious wounds they would often bandage the packed wound really tightly to increase the irritation, and would arrange a goose quill (or some other tube), emerging from the dressing, to act as an outlet pipe to drain off the pus as it formed.

By contrast, in the case of treatment by the weapon salve, the wound would simply be kept clean and dry, perhaps with a light dressing to keep it clean, while the doctors concentrated on smearing their obnoxious potions on a sword or a blood-stained cloth. It's easy to see, therefore, why the weapon salve was so successful as a cure for wounds. The experimental evidence was on the side of magic. The theory of laudable pus was so strongly held that on-lookers concluded *not* that the theory of laudable pus must be wrong, but that somehow the weapon salve had an occult power of curing wounds without the formation of pus. It took a long time to sort this out and arrive at the real truth of the matter.

The process of the amalgamation of magic into natural philosophy to form a new, reformed, philosophy of nature was bedevilled (if we can put it that way) by many complications of this sort. Experimental evidence does not speak for itself, but must be interpreted. Unfortunately, interpretations tend to fit into preconceived expectations and assumptions, and often obscure alternative possibilities. Francis Bacon was fully aware of these difficulties with experimentalism and tried to overcome, or at least minimise, them in the detailed working out of his method. Even if not entirely successful, it is this aspect of Bacon's enterprise that made it so valuable in the subsequent development of science. Bacon clearly showed the importance of thinking not just about the content of one's science, but also of its method. Thinking about how knowledge claims were arrived at helps you to determine the validity and strength of those claims.

In the end, then, much of natural magic was combined with many aspects of natural philosophy to give rise to something closer to modern science. But something else happened during this process. Magic had always had what today would be called 'a bad press'. The religious authorities, whether the Roman Catholic Church or later one or other of the Protestant Churches, always resented the practice of magic. Its demonological aspects were always seen as blasphemous and heretical in the extreme, but even natural magic was seen as a threat to sound religion.

As Cornelius Agrippa pointed out, magicians often accomplish 'things that are popularly held to be miracles when they are really no more than anticipations of natural operations' (*Of the Uncertainty and Vanity of the Sciences*). But anything that raised confusion in the popular mind as to what was a miracle and what might have been accomplished by natural means was seen as threatening to the Churches. Nobody wanted to risk drawing the anger of the Church. Consequently, magic was always denounced and denied, even by those who were clearly exploiting it. As we've seen, for example, Bacon was always keen to show his contempt for magic and to dismiss its claims.

What this meant, therefore, was that during the process of amalgamation of magic into natural philosophy, it was seldom, if ever, acknowledged that magical ideas were being appropriated. The good ideas in magic were silently incorporated into the reformed natural knowledge, while the bad ideas were used, more and more vigorously, to denounce magic as a sink of false and ludicrous beliefs. Demonology, once nothing more than a desperate means of acquiring knowledge of the natural powers of things, became a much more substantial element in the rump of magic that was left after the most fruitful aspects of magic had been creamed off into the new philosophy. Similarly, astrology, once taught in every medical school in Europe as an essential element in medical diagnosis and prognosis, was dropped from the medical curriculum and became another example of the ridiculousness of magical claims.

Magic became so removed from anything that might be considered to be naturalistic that the very term 'natural magic' dropped from view. In the increasingly secular world of the eighteenth century, the term 'supernatural' came to be applied loosely to anything that wasn't obviously physical and materialistic, and so magic increasingly came to be seen as concerned with a supernatural realm in which only the superstitious believed.

Historians of science who have said that magic cannot have played an important part in the development of modern science are obviously thinking of magic in these post-eighteenth-century terms, and are extrapolating it back to the sixteenth and early seventeenth centuries. But this is clearly anachronistic. If we want to assess whether Bacon or Newton (or any other figure) might have been influenced by magical ideas and beliefs, we need to know what they considered magic to consist of, and therefore what they might have believed, and not foist our own ideas of magic on to them.

Sir David Brewster, a scientist in his own right but also the first biographer of Isaac Newton, famously expressed his dismay that Newton, in his alchemical notes, should waste his time considering writings by an alchemist who was obviously (to Brewster, anyway) a 'knave and a fool'. You might imagine that one who admired Newton as much as Brewster clearly did would take Newton's interest in alchemy as a sign that there must have been something in it, or that there must have been good reasons in

Newton's day for thinking there was something in it. But, no. Brewster knew that alchemy in his day was meaningless mumbo-jumbo, and so Newton's belief in it must have been nothing less than a lamentable aberration.

The fact is that Isaac Newton, working at the end of the seventeenth century, accepted many ideas from the natural magic tradition. What's more, although Newton has been described as 'the last of the Magi', he was not alone in his beliefs. Similarly, at the beginning of the century, Francis Bacon was not alone in looking to the natural magic tradition for new insights into the operations of nature. Nor was he alone in believing that the traditional natural philosophy of the universities was immured in an inward-looking scholasticism, which was making no new discoveries, and had never done anything for the public good and the amelioration of the human condition. Nor was he the only one to notice the recent successes of magically inclined thinkers like Paracelsus or William Gilbert, and the increased intellectual respectability of magic after the discovery of the ancient magical wisdom of Hermes Trismegistus.

Bacon was, however, the only one invested with such a strong sense of mission that he was destined to serve mankind by doing something that would prove universally useful, that he decided to reform the system and the method of natural philosophy. There can be no doubt that, in order to help him fulfil that mission, which he believed could best be done by ensuring that 'human Knowledge and human Power, meet in one' (*New Organon*, I, Aphorism 3), Bacon turned to the traditions of magic.

• Chapter 8 •

Separating Science and Religion?

If there has been a tendency among historians to play down Bacon's appropriation of magical ideas, there has been an equal effort to play down the role of religion in shaping his life's work. Indeed, there have been a number of attempts ever since the eighteenth century (when the secularisation of Western culture began in earnest) to see Bacon, if not as one of the first atheists, at least as one of the first to show a healthy lack of interest in religious considerations. We can see the beginnings of this in the Enlightenment of the eighteenth century, when intellectuals saw the recent triumphs of the new philosophy, especially Newtonianism, as the result of its secure rational and experimental foundations, and began to believe that the scientific method could be used to establish moral and political truths.

For Enlightenment thinkers, scientific knowledge was seen as an intellectual authority capable of replacing religion. They took their cue from the closing remarks in Isaac Newton's second great book, entitled *Opticks* (1704), that 'if natural

Philosophy in all its parts, by pursuing this Method, shall at length be perfected, the Bounds of Moral Philosophy will be also enlarged'. Accordingly, Enlightenment thinkers conceived 'human sciences' (political economy, economics, psychology and so on), the principles of which could be established by following the scientific method. In morals, for example, principles of good and evil could be established not just by accepting what it says in the Bible, but by following rational principles. Murder, theft and much lesser crimes could all be shown to be incompatible with the basic commandment for life in a civic society: 'Do unto others what you would have others do unto you.'

Given this background, Enlightenment thinkers wanted to see the heroic figures in the history of the new science as thinkers swayed only by rational and empirically grounded principles. As far as the secularist philosophers of the Enlightenment were concerned, this meant their heroes must have been at least indifferent to the superstitious doctrines of religion, and preferably nothing more than deists or even atheists. A deist was someone who accepted the existence of God on the grounds that the natural world was too well organised, too intricately designed and too interdependent to have come about without a divine creator, but who nevertheless rejected the teachings of the organised religions. A deist, therefore, would believe in a creator God, but not in Christ, virgin births, miracles, bodily resurrections from the dead and so on. Since Bacon was one of the greatest heroes of Enlightenment thinkers, there was a strong tendency for them to see him as a deist or an atheist.

Bearing in mind that irreligion was still a capital offence during his time, Enlightenment thinkers knew that Bacon could not have announced that he disbelieved in revealed religion. Some were convinced, however, that he could have left sufficient hints or 'coded' allusions in his writings to indicate his real views to those who knew how to read between the lines. (These thinkers were convinced of this for the simple reason that they themselves played the same game in their own writings.) Certainly there are a number of instances in Bacon's writings that can easily be taken this way. Consider the story he uses to illustrate the nature of superstition:

And therefore it was a good answer that was made by one who when they showed him hanging in a temple a picture of those who had paid their vows as having escaped shipwreck, and would have him say whether he did not acknowledge the power of the gods – 'Aye,' asked he again, 'but where are they painted that were drowned after their vows?' (New Organon, *I, Aphorism 46*)

Another coded support for atheism was found, naturally enough, in Bacon's essay 'Of Atheism'. At one point Bacon insists that the standard Aristotelian philosophy taught in the universities, which explains all physical phenomena in terms of the interactions of four elements (earth, water, air and fire), is more likely to provide support for atheism than the atomist philosophy, which explains all phenomena in terms of the intermingling and incessant motions of virtually infinite numbers of invisibly small atoms. It is more likely, Bacon explains, that only four elements might randomly interact in the appropriate way than that an infinite army of atoms should randomly give rise to the world around us.

On the face of it, the discussion reads like a defence of atomism against charges of atheism. Enlightenment thinkers turned the story on its head, however. Since atomism had always been associated with atheism, Bacon's insistence that Aristotelianism (which had been adopted as the official natural philosophy of the Roman Catholic Church in the Middle Ages) provided better support for atheism than atomism was a clear signal that atheism, no matter which natural philosophy you subscribe to, must be true.

But the claim of Bacon's that has often been presented as most irreligious in its import is his insistence that natural philosophy and religion should always be kept separate. Bacon insisted that the study of the natural world had been blighted by 'a troublesome and intractable enemy' – namely, 'blind immoderate religious zeal'. This could be seen, he explained, not only in ancient times, but also at present:

> *Among the Greeks those who first suggested to men's untutored minds that thunderbolts and storms had natural causes were condemned for impiety. On the accusation of some of the early Christian fathers Cosmographers, who on clear evidence which no sane man could reject today, claimed that the earth was a sphere and therefore inhabited at the antipodes, fared little better than the Greeks. They were brought to trial for impiety … In our own days discussions concerning nature have been subjected to even harsher constraint (Thoughts and Conclusions).*

Indeed, if anything the situation was getting worse, according to Bacon, because there were strenuous efforts to 'celebrate a legal marriage between Theology and Natural Philosophy, that is between Faith and the evidence of the senses', so that only those scientific ideas compatible with theology were allowed. The result, as far as Bacon was concerned, could only be 'a disastrous confusion between the human and the divine' (*Thoughts and Conclusions*).

Similarly, in his *Advancement of Learning* he wrote that when natural and divine philosophy were 'commixed together' it could only lead to 'an heretical religion and an imaginary and fabulous natural philosophy'. Science and religion, he seems to be saying, don't mix.

Secular and anti-religious writers have seized upon such comments to present Bacon as one of their own: an essentially secular thinker with little regard for the superstitious claims of religious believers. A slightly less blinkered look at what Bacon actually had to say about the mixing of science and religion will quickly reveal, however, that these commentators, ever since the Enlightenment, have simply been reading Bacon by their own lights. Contrary to their interpretation, Bacon was not so much concerned that science and religion should not be mixed, but that they should not be mixed in the wrong way.

We cannot understand Bacon's real concern about the inter-relationship between science and religion by comparing his statements with our own secularised views on the matter. We need to consider his statements in the light of the historical context within which he lived and thought. Bacon died before the public scandal of the Galileo affair (when in 1632 the famous Italian mathematician, Galileo Galilei, was condemned by the Catholic Church for promoting the new idea that the Earth orbited around the Sun, instead of remaining stationary at the centre of the universe), and long before Charles Darwin's *Origin of Species* (1859) caused another major rift between science and religion.

Bacon did not live in a secular age, and all his mental instincts would have led him to think in devoutly Christian terms. Accordingly, the 'rightful station' of natural philosophy, he always believed, was 'as the accepted and loyal handmaid of religion'. 'For religion reveals the will of God', he explained, and 'natural philosophy His power' (*Thoughts and Conclusions*).

The presumption that natural philosophy should act as the handmaid of religion was an entirely traditional view, dating from at least the thirteenth century, and until the sixteenth century had persisted without significant controversy. But the thirteenth to the early sixteenth century had been a time when there was only one religion in Western Europe, Roman Catholicism. Things changed in the sixteenth century with the Protestant Reformation and the establishment of the major Protestant Churches, Lutheranism and Calvinism, and various lesser Protestant sects, each of which rejected the corrupt Roman Church, and presented themselves as the one true Church.

Given the traditional view that natural philosophy provided support and confirmation for the religious doctrines of the Church, it was inevitable that each of the new Churches would wish to show how correct natural philosophy supported their doctrines and contradicted the doctrines of another Church. Where once there had been a uniform natural philosophy to defend a unified Church, there was now a splintered natural philosophy to defend a fragmented Church.

Bacon was by no means the only natural philosopher to believe that the result was an unfounded distortion of natural philosophy to fit the required religious doctrines, or a compromising of religion in an attempt to capitalise on supposed scientific truths, or both. Instead of settling religious disagreement and deciding between rival doctrines, natural philosophy itself was torn apart as different interpretations were imposed on its evidence.

For example, the Roman Catholic Church had recently pronounced the immortality of the soul to be an official doctrine, but Martin Luther, the initiator of the Protestant Reformation, believed this to be a pagan belief, not a Christian one. Indeed, it is almost certainly true that the Papacy made the immortal soul part of official dogma at this time thanks to the influence of the newly discovered writings of the Ancient Greek philosopher Plato (*c.* 427–347 BC) and various Neoplatonic works such as those attributed to Hermes Trismegistus.

Luther was perfectly correct in insisting that all the talk of immortality in the Bible concerns the resurrection of the body; there is no clear scriptural warrant for a belief in the immortality of the soul (you can check this out for yourselves). Here, then, was a clear case where, many believed, natural philosophy might be used to settle the matter. Accordingly, Roman Catholics developed various arguments to establish that the soul must be *naturally* immortal. The very nature of the soul meant that it had to be immortal. Lutherans, by contrast, insisted that there was nothing about the nature of the soul that made it (or could make it) immortal. The soul could die with the body, therefore, and be reincorporated on the Day of Judgement, when all the dead were to be bodily resurrected.

This dispute was driven by the doctrinal division between Catholics and Protestants, but others were driven by commitment to natural philosophical positions. Bacon himself seems to have been most concerned about the efforts of Paracelsians to interpret the Bible in accordance with their alchemical views. Bacon subscribed to an understanding of the world picture which was largely inspired by Paracelsian views, and this presumably accounts for his particular anxiety in seeing these views extended to Biblical interpretation and, by implication, religious teachings. Paracelsus was aptly known as the 'Luther of medicine'. Although this epithet was primarily intended to imply that his theories were as disruptive in medicine as Luther's were in religion, it also points to the very real fact that Paracelsus saw himself as a religious reformer.

The Paracelsians, concerned primarily with the

medicinal usefulness of alchemy (rather than with converting lead into gold), justified their medical theories by claiming that the human body worked by alchemical processes. The digestion of food, respiration and other bodily functions were seen as precisely analogous to procedures carried out in retorts, alembics, furnaces and so on by alchemists. Since the Paracelsians also believed in the sympathetic correspondence between the so-called microcosm and the macrocosm (see Figure 4) – that is to say, between the 'little world' of man and the greater world – they wanted to show that God created the world as a whole, not just man, by using alchemical processes.

Consequently, the Paracelsians gave an alchemical interpretation of the description of the Creation in the opening chapters of the Book of Genesis. When 'the Spirit of God moved upon the face of the waters' in verse 2 of chapter 1, for example, the Paracelsians saw God as an alchemist separating the waters that were above the firmament from the waters below the firmament (verse 7) by a process of cosmic distillation. Bacon expressed his disapproval clearly: 'Some of the moderns have with extreme levity indulged so far as to attempt to found a system of natural philosophy on the first chapter of Genesis', he noted, but 'from this unwholesome mixture of things human and divine there arises not only a fantastic philosophy but also an heretical religion' (*New Organon*, I, Aphorism 65).

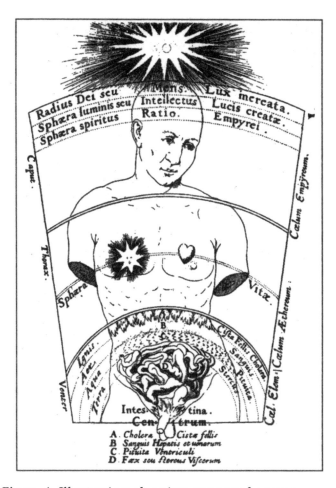

Figure 4: Illustration showing supposed correspondences between the microcosm, man, and the macrocosm, or world as a whole. This was an ancient and widespread belief, but was especially prominent in the magical tradition. The illustration is from the second volume of Robert Fludd's *Utriusque Cosmi Historia* (*History of Both Cosmoses*, Oppenheim, 1619).

As far as Bacon was concerned, natural knowledge was not yet certain and secure enough to be used safely and unequivocally in establishing religious truths. Bacon might well have felt that Paracelsian alchemy was a better bet than any of its rivals in natural philosophy, but he was too convinced of the truth of the moderate Calvinism of the Elizabethan Church of England to subscribe to Paracelsian religious views. As far as Bacon was concerned, therefore, natural philosophy could only suffer by being dragged into intractable religious controversy, such as the dispute over the immortality or mortality of the soul. And religion could only lose if it tried to derive its principles not from the revelations of Holy Scripture, but from the partisan claims of natural philosophers like the Paracelsians.

It was better, he therefore believed, to keep science and religion apart, but this certainly did not mean that he felt religion was an irrelevance. On the contrary, he clung to the traditional view that the correct natural philosophy, if we could discover it, would be the perfect handmaid to the true religion. By the time he was writing these things, 'Religious controversies have become a weariness of the spirit', he said, and it was perhaps better to 'contemplate the power, wisdom and goodness of God in His works'. But this should only serve to counter irreligion, not to promote a particular faction of the Christian Church. Caution is the watchword.

> *To conclude, therefore, let no man, upon a weak conceit of sobriety or an ill-applied moderation, think or maintain that a man can search too far or be too well studied in the book of God's word [scripture] or in the book of God's works [nature]; divinity or philosophy; but rather let men endeavour an endless progress or proficience in both; only let men beware ... that they do not unwisely mingle or confound these learnings together* (Advancement of Learning).

CHAPTER 9

RESTORING LOST DOMINION:
BACON AND THE MILLENNIUM

Bacon's belief in the importance of natural philosophy as a handmaid to science did not merely appear in his works as a pious reaffirmation of a long-standing tradition in natural philosophy. He developed the idea in a unique and what was to prove a highly influential way. Bacon's religious devotion was so strong that it would be surprising if he did not seek a religious justification for his ambition to extend 'the bounds of Human Empire to the effecting of all things possible' *(The New Atlantis,* 1627). It just so happened that Bacon was working at a time when there was a huge revival of millenarianism in England, and he immediately saw himself as engaged upon a millenarian enterprise.

Millenarianism, or millennialism, was the belief that the Second Coming of Christ, upon the Day of Judgement, was about to take place. The names for this belief derive from the concomitant belief that the Second Coming was associated with a thousand-year rule of the chosen people (in the Christian tradition, Christians, of course, but they also had to be of the right faction). Based on close and highly inventive readings of the so-called apocalyptic books of the Bible (chiefly Daniel in the Old Testament and The Apocalypse, or Revelation of St John, in the New), and their supposed match with the history of Europe and the Holy Land since the days of the Apostles, some believed the millennium had already run its course and was about to be ended by the Second Coming, while others believed it was about to begin with the imminent Second Coming. Either way, the culminating point of Christian history was almost ready to take place.

We noted earlier that these beliefs can be seen at work in shaping Bacon's beliefs about his destiny and his ambitions for his life's work. He saw his *Great Instauration*, his comprehensive revision of all natural knowledge, as a fulfilment of Daniel's prophecy that 'science will advance' in the last days of the world, and as a preparation for the ultimate Sabbath, the Day of Judgement, at the end of the world. The religious importance of his work could be seen, Bacon believed, in the fact that it would repair the loss of dominion over nature that mankind lost at the Fall, after Adam and Eve ate the forbidden fruit. In the closing words of the *New Organon*, Bacon clearly indicated that his desire to produce 'an improvement in man's estate, and an enlargement of his power over nature' was linked to his belief that the consequences of the Fall could be reversed:

For man by the Fall fell at the same time from his
state of innocency and from his dominion over
Creation. Both of these losses however can even in
this life be in some part repaired; the former by
religion and faith, the latter by arts and sciences.
For Creation was not by the curse made altogether
and forever a rebel, but in virtue of that charter,
'In the sweat of thy face shalt thou eat bread'.

Describing his own work in the third person, in
his *Thoughts and Conclusions*, Bacon wrote again in
terms of religious auspices.

Bacon consulted the auspices with all due care; and
here the first thing that struck him was that the
business in hand, being eminently good, was mani-
festly of God, and in the works of His hand small
beginnings draw after them great ends. Then the
omens from the nature of the Time were also good.

To illustrate what he means by the nature of the
Time, and its favourable omens, he alludes once
again to the prophecy in Daniel about the last days.

It ought not to go for nothing that through the
long voyages and travels which are the mark of
our age many things in nature have been revealed
which might throw new light on natural philos-
ophy. Nay, it would be a disgrace for mankind if

the expanse of the material globe, the lands, the seas, the stars, was opened up and brought to light, while in contrast with this enormous expansion, the bounds of the intellectual globe should be restricted to what was known to the Ancients.

Writing like an apocalyptic scholar, seeking to persuade his readers that the end of history is nigh, Bacon sums up the history of the advance of science:

Suppose we allow twenty-five centuries to the recorded history of mankind. Of these scarce five can be set apart as propitious towards, and fruitful in, scientific progress, and the kind of sciences they cultivated were as far as possible removed from that natural philosophy we have in mind. Three periods only can be counted when the wheel of knowledge really turned: one among the Greeks, the second with the Romans, the last among the nations of Western Europe. All other ages have been given over to wars or other pursuits. So far as any scientific harvest is concerned they were barren wastes.

The fact that things are now very different, and natural philosophy is progressing, shows that the last days are coming, or so Bacon wants to imply.

There can be little doubt that Bacon frequently and unselfconsciously thought about his proposed reform of natural knowledge in millenarian terms. In his *Refutation of Philosophies* he even invoked one of the major characters from the apocalyptic vision of the last days in the book of Revelation, the Antichrist. The chief enemy of Christ, the Antichrist figured largely in apocalyptic literature. Mentioned in the First Epistle of John as a sign by which 'we know that it is the last time' (I John, 2, 18), and playing a prominent role in Revelation, the Antichrist was often identified by Protestant writers as one of the Popes (Boniface VIII and John XXII were favourites) or simply with the Papacy in general. Bacon himself was not directly concerned to provide his own apocalyptic inter-pretations but, as we've seen, he did tend to see his own enterprise as part of the fulfilment of the prophecies of the last times. It is hardly surprising, therefore, that at one point he likens (but doesn't identify) the Ancient natural philosopher Aristotle to the Antichrist.

Aristotle must rank as one of the most influen-tial philosophers of all time. His system of natural philosophy held complete sway throughout the Middle Ages and into the early modern period, from the thirteenth century to the beginning of the seventeenth. By the late sixteenth century, however, it was becoming increasingly clear to leading thinkers that his natural philosophy was not all it was cracked up to be. By the beginning of the seventeenth century Bacon was by no means alone in seeking to reject the Aristotelian system.

The trouble was, however, that Aristotle's system was so comprehensive in its coverage, and so cunningly wrought, that it was extremely difficult to replace in a piece-meal fashion. Aspects of Aristotelianism had recently been shown to be wanting, notably his earth-centred cosmology, in which the Sun supposedly moved around the Earth along with the Moon and planets. But since Aristotle's cosmology was tied up with his physics and his theory of motion, it was hardly possible to accept the alternative moving-earth cosmology (developed by the Polish astronomer, Nicolaus Copernicus) without first developing a new theory of motion and a new physics.

In Italy, Galileo took up the challenge of trying to develop a new, non-Aristotelian theory of motion. However, his efforts only raised further questions relating to other connected but unrevised aspects of the Aristotelian system. What was needed was a completely new system of natural philosophy that was capable of replacing Aristotelianism lock, stock and barrel. In a very real sense, this is what Bacon saw himself as trying to do in the Great Instauration, and, although he did not achieve this in his lifetime, it could be said that his influence was such that his suggested reforms did eventually achieve the complete overthrow of Aristotle.

It was while dismissing the authority of Aristotle that Bacon likened him to the Antichrist.

There is a question we should put to ourselves. Does the fact that Aristotle drew to himself both earlier and later ages prove him truly great? Oh, great without a doubt; but no greater than the greatest of impostors. For this is the prerogative of imposture, and in especial of the Prince of imposture, the Anti-Christ. 'I am come', says the Truth Himself, 'in the name of my Father, and ye do not receive me; but, if one cometh in his own name, him ye will receive' [John 5, 43] ... Christ says that he who comes in the name of the Father... will not be received; but he who ... usurps authority to himself and comes in his own name, him men will follow. Now if any man in philosophy ever came in his own name, Aristotle is that man (Refutation of Philosophies).

It seems highly likely from this quotation that Bacon was one of those Protestant thinkers who dismissed Aristotle as a pagan who had gained his authority from the Roman Church. Aristotelianism had been embraced by the Catholic Church since the thirteenth century as the official 'handmaid' to its religion, and Aristotle's unchallenged authority throughout the Middle Ages and into the sixteenth century was due in no small measure to his affiliation to the Mother Church. Small wonder that an English Protestant like Bacon, entertaining apocalyptic thoughts, should link Aristotle, the natural philosopher of Catholicism, to the Antichrist so often identified by Protestants with the Papacy.

Besides, as a handmaid to religion, Aristotelianism had proved lamentable, in Bacon's eyes, doing nothing to restore man's lost dominion over nature, but merely showing university-trained philosophers how to spin useless cobwebs of learning out of their own minds. Certainly, it was usual in the universities for the professors of philosophy, and their students, to answer all questions about the nature of the physical world, not by studying the world itself, but by carefully scrutinising what Aristotle said on the matter, and considering what he might have meant. Cobwebs of learning seems a very apt image.

Bacon's new natural philosophy, by contrast, is to usher in Biblical prophecies of advancing science and thereby act as a useful handmaid to religion. Bacon's natural philosophy, properly pursued, will prove to be 'after the word of God at once the surest medicine against superstition, and the most approved nourishment for faith, and therefore she is rightly given to religion as her most faithful handmaid' (*New Organon*, I, Aphorism 89). As the instigator of this new system of natural philosophy, Bacon prays to the Holy Trinity, in the Preface to the *Great Instauration*, so that 'they will vouchsafe through my hands to endow the human family with new mercies'. In this way, the human race will 'recover that right over nature which belongs to it by divine bequest' (*New Organon*, I, Aphorism 129), but which was lost after the Fall.

Bacon was writing at a time when Protestants were increasingly concerned about the historical accuracy of the apocalyptic prophecies. Where once the descriptions of events leading to the Second Coming and the Day of Judgement were seen as merely allegorical tales of good and evil, and the difficulties of being able to distinguish between genuine good and the deceptions of the Devil or the Antichrist, Protestants began to see them as accounts of the past, present and future of the Church. After first being doubtful about the authenticity of the book of Revelation, Martin Luther realised that ingenious interpretation of its prophecies could be used to support the Lutheran message.

Where Luther once saw Antichrist as a merely literary representation of evil and opposition to Christianity, by the 1530s he was claiming confirmation of his views about the corruption of the Papacy in the fact that 'history agrees with Scripture ... that the Pope is Antichrist'. By showing how predictions made in apostolic times came true in the Reformation, Luther and others were able to show that they were bringing about the Time of the End, establishing the True Church and ushering in the Second Coming of Christ.

Showing how *some* of the strange prophetic visions of Revelation, correctly interpreted, precisely coincided with actual historical developments, or with events taking place in the present, established the credentials of the millenarian interpreters. They could then use their interpretations of the rest of the visions to persuade their readers what they must do to bring about the predicted destiny of the Church. The trick was all in the interpretation, of course. The more plausible the accounts of what or who was meant by Gog and Magog, the Whore of Babylon, the woman clothed with the Sun, the red dragon with seven heads and ten horns, the beast whose number was 666 and all the rest of the cast in the Apocalypse, the more persuasive one seemed as an interpreter of God's words. Needless to say, this proved difficult and controversial. Bacon himself included prophetic history as one of the deficient areas of learning that needed to be pursued 'with wisdom, sobriety and reverence' in order to bring about the *Advancement of Learning*.

History of Prophecy, consisteth of two relatives, the prophecy and the accomplishment; and there-fore the nature of such a work ought to be, that every prophecy of the Scripture be sorted with the event fulfilling the same, throughout the ages of the world; both for the better confirmation of the faith, and for the better illumination of the Church touching those prophecies which are yet unfulfilled (Advancement of Learning).

It can hardly be a coincidence that this follows just a few sentences after one of the several places in Bacon's writings where he presents his own interpretation of Daniel 12, 4: 'Many shall run to and fro, and knowledge shall be increased.'

Whatever we might think about the efforts of thinkers like Bacon and his contemporaries to discover the correct interpretations of the Bible, and to confirm those interpretations by showing how they fit with historical fact, they are clearly revealing about the nature of Bacon and his work. Far from being a covert atheist, as was once thought (and, indeed, as some modern authors on Bacon still like to think), a correct interpretation of his writings shows that, to a large extent, his enterprise was shaped and driven by a firmly held belief about one's Christian duty. Bacon believed that by hard work – '[I]f we labour in thy works with the sweat of our brows' ('Plan of the Work', *Great Instauration*) – it should be possible to repair the loss of man's dominion over Creation. And we know it should be possible, he would have insisted, since it seems to be linked to man's ultimate destiny, as predicted in scripture about the 'time of the end'.

It is important to note that Bacon should not be included among the radical sectarian believers in a coming apocalypse. Scholars have tended to dismiss Bacon's millenarianism in the belief that all millenarians were politically subversive radicals, hell-bent (if that's the right phrase) on bringing an end to the current order of society and replacing it with a system of Christian communism. Bacon's politics were entirely concerned with upholding the *status quo*, and his religious views were very definitely moderate. Nevertheless, it was possible to be a moderate yet still believe that Jesus would come again one day, and that one's Christian duty was to prepare quietly for that day. As the poet John Milton so famously put it in his poem 'On His Blindness', 'They also serve who only stand and wait'. There need be no doubt, therefore, of Bacon's sincerity when he wrote:

It is not the pleasure of curiosity, nor the quiet of resolution, nor the raising of the spirit, nor victory of wit, nor faculty of speech, nor lucre of profession, nor ambition of honour or fame, nor inablement for business, that are the true ends of knowledge; some of these being more worthy than other, though all inferior and degenerate: but it is a restitution and reinvesting (in great part) of man to the sovereignty and power ... which he had in his first state of creation (Valerius Terminus, *1603).*

• Chapter 10 •

Building a Model of the World: Bacon and Utopia

To the secular mind, the idea of the end of the world seems appalling – certainly not something to be longed for. But it's very different if you have faith in God and a firm belief that you are one of His chosen ones. The seventeenth-century believers who tried to work out the time of the Second Coming had a strong yearning for it to happen; the sooner the better. Their vision of the end of all things did not entail their own obliteration, merely an end to their former way of life.

Descriptions of the state of affairs after the Second Coming of Christ read like attempts to describe life under a perfect political system, where all are equally happy as social equals, ruled over by a supremely benevolent, just and completely incorruptible ruler (Jesus Christ). The vision of social and political perfection was also helped by the assumption that everyone shared the same values and beliefs. The Day of Judgement was a day of what would now be called religious and ethnic cleansing. All Catholics, or all Protestants (depending on your own persuasion), all heretics, dissenters and sceptics, all Jews, all Muslims and everyone different in any way that could possibly be disapproved of, were consigned to Hell.

These futuristic fantasies were never concerned with spiritual, disembodied or unrecognisable mystical states. Everyone's alternative reality might have qualified as alternative, but it was still recognisably like reality: human bodies, living and interacting together, with all that that entailed, including systems of moral prescriptions and government. Some millennial expectations were pastoral, others were urban; not one of them seemed to be concerned to escape the material world. It is hardly surprising, therefore, that such increased hankerings after earthly happiness, in the apocalyptic fervour of the sixteenth and seventeenth centuries, should lead to a burgeoning of utopian fantasies in contemporary literature. Utopian visions turned a yearning for states of bliss into a dream of a blissful state.

Utopian literature has a long history, with some notable examples dating from Ancient times. But it seems entirely fitting that the genre should take its name from a superb example written in the sixteenth century. Thomas More's *Utopia* (1516) initiated what might be called the golden age of utopian literature. Apart from the stimulus provided by millennial fervour, there was the example set by the newly rediscovered *Republic*, written by the much-admired Ancient philosopher, Plato, describing his ideal city state. Furthermore, this was an age that began to be aware of cultural relativism.

The voyages of discovery that made such an impression on Bacon stimulated a literature of travel, often describing the wonders of far-off civilisations and their lifestyles. It was also an age of severe social problems arising from changes in the political economy of Europe. As the traditional feudal system came to an end, landowners dispossessed subsistence farmers of their lands to make way for animal pasture, and the success of plundering new colonies abroad caused massive monetary inflation at home. The social conditions that made the common man yearn for the Second Coming of Christ made intellectuals speculate about what was needed to make the perfect society.

Bacon, Lord Chancellor to James I (as More was Lord Chancellor to Henry VIII), also wrote a utopian fantasy. It is easy to understand why. Believing, as he did, that his reforms in the method of gathering and confirming knowledge of the natural world would enable man to exploit natural phenomena for the relief of man's estate, and cause a general improvement in the conditions and the experience of life, it seems natural that he should try to show how. In fact, Bacon's utopia, *The New Atlantis*, was written in 1624, towards the end of his life, and was published posthumously in 1627 as a companion piece to what was nothing more than another sample part of Bacon's unfinished *Great Instauration*, the *Sylva Sylvarum* (or *Forest of Forests*).

The culminating part of the *Great Instauration* was to be a 'New Philosophy' or, as Bacon also called it, 'Active Science'. This is how Bacon described it in his 'Plan of the Work', which in fact was nothing more than an announcement and outline of the whole scheme:

The sixth part of my work for which the rest are but the preparation, will reveal the philosophy which is the product of that legitimate, chaste, and severe mode of enquiry which I have taught and prepared. But to perfect this last part is a thing both above my strength and beyond my expectation. What I have been able to do is to give it, as I hope, a not contemptible start. The destiny of the human race will supply the issue, and that issue will perhaps be such as men in the present state of their fortunes and their understandings cannot easily grasp or measure. For what is at stake is not merely a mental satisfaction but the very reality of man's wellbeing, and all his power of action.

Given the extent of the preparative part of his work – for example, the gathering of the forest of forests of information required for the encyclopaedic database – it is hardly surprising that Bacon saw the culmination of his scheme as something beyond his strength and his expectation. But between 1620 and 1624, he evidently hit upon the brilliant idea of describing a utopia to illustrate what 'men in the present state of their fortunes ... cannot easily grasp' (*The New Atlantis*), about the benefits to their wellbeing which might follow from his project. *The New Atlantis* was published, therefore, together with the *Sylva Sylvarum*, just after his death.

It's possible that Bacon took the idea for *The New Atlantis* from two recently written utopian works that also saw natural magic and millenarian religion as important elements in the ideal societies they described. It's not certain whether Bacon knew either of these works, but he could easily have been aware of them, even if he hadn't actually read them. There is no denying, anyway, thanks to their emphasis upon the importance of natural knowledge, that Bacon's *New Atlantis* can be lumped together with Tommaso Campanella's *Civitas Solis* (*The City of the Sun*, 1623, but written in 1602), and Johann Valentin Andreae's *Christianopolis* (*Reipublicae Christianopolitanae Descriptio*, 1619).

Bacon certainly knew of Campanella, who was, like him, a great admirer of another Italian philosopher, Bernardino Telesio. It's possible, therefore, that Bacon kept an eye on Campanella and his works, knowing him to be a close disciple of Telesio's. Campanella himself was every inch the

Renaissance magus, being especially committed to what was known as stellar magic. Based on the astrological assumption that the heavenly bodies somehow influenced events and processes on the Earth, the stellar magician was not content merely to predict what the influences of the heavens would be: he sought to change them. If, therefore, your horoscope foretold disaster, what you needed was a stellar magician who, by using the influences of Earthly things corresponding to the heavenly bodies, could counteract the heavenly tidings (remember John Donne's poetical comment about the Artist who could 'constellate anything' by imprisoning the influence of the stars 'in an Hearbe, or Charme, or Tree' and 'do by touch, all which those stars could do').

Campanella's reputation as a stellar magician was so great that he was finally sprung from the dungeons of the Inquisition, where he had been imprisoned for nearly 30 years, by the Pope himself, Urban VIII, whose grisly demise was being astrologically predicted by his many enemies. The Pope certainly felt himself in need of a stellar magician.

Campanella's fictional *City of the Sun* was built on a large hill and was organised in seven concentric zones, mounting up the hill and divided from one another by seven circular defensive walls. It was modelled on the concentric heavenly spheres of the Sun and six known planets, and was significant for embodying the arrangement as described by Copernicus (with the Sun at the centre and the Earth in motion around it with the other planets), which was itself a radically new and still not accepted idea.

The walls, both sides, were used to record all the knowledge of the world and were used in the education of children (whose classes moved around each of the walls and, successively, from outer to inner walls). The nature and qualities of minerals were recorded on the inside of the second wall, for example, and a sample of each was embedded in the wall. Similarly, on the outside of the same wall, different liquids were described and samples were provided in small carafes built into the wall. The order of subjects followed the order of the Great Chain of Being and in all cases the descriptions of the different things included their 'correspondence to celestial and earthly things'.

Astrology played a major role in the lives of the Solarians, as the inhabitants were called. The rhythms of life in the city, the timing of daily routines, as well as more significant events such as the sowing of seeds and harvesting, were dictated by the determinations of the astrologers.

Although this may look like superstitious nonsense to us, it is important to remember that astrology was by no means discredited in Campanella's and Bacon's day. It was considered to be so useful that it was the only one of the magical arts to be taught in the universities. Astronomy was taught as a necessary part of one's training to become a Master of Arts (in those days, the first degree), so that, if you wished to go on to train as a medical doctor, you could progress to the more demanding art of astrology, interpreting the significance and likely influence upon patients of astronomical configurations. Everyone took it for granted that the stars affected your health, so it was hardly surprising that Campanella should see knowledge of the stars and their influence as an important element in maintaining an ideal society.

Given Bacon's own beliefs about the pragmatic superiority of natural magic over scholastic natural philosophy, we can be absolutely sure that if he ever did see Campanella's *City of the Sun*, either while it was circulating in manuscript or after it appeared in print in 1623, he would not have regarded it as superstitious, but would have recognised it as a utopia based on exploiting knowledge of nature for the benefit of the inhabitants.

Precisely the same can be said about Andreae's *Christianopolis*. Andreae was an alchemist in the mould of Paracelsus or Van Helmont. That is to say, he was one of those who subscribed to an entire cosmology based on alchemical assumptions. For Andreae, the processes of the natural world, in the macrocosm and the microcosm, were all alchemical processes. In one of his earlier writings, *Fama*

Fraternitatis (*Report of the Brotherhood*, published 1614 but written before 1610), Andreae included this new breed of alchemist among the recent gifts bestowed on man by God since the Reformation.

> *In as much as the wise and kind God has lately poured His mercy and kindness over the human race, our understanding of the universe and His Son has constantly increased and we have good reason to boast of a fortunate time. Not only has He uncovered for us half of the unknown and obscure worlds, but He has also given us many strange creatures and occurrences that have never before happened. He has also let enlightened spirits arise who are to restore the somewhat maligned and imperfect art of alchemy to its rightful place, so that man will finally discover his dignity and glory and recognise how he is but a microcosm and how deep his art is rooted in nature.*

If you think about them, Andreae's views here are very close to Bacon's when he comments on his favourite part of the book of Daniel. Andreae refers to the voyages of discovery ('[God has] uncovered for us half of the unknown and obscure worlds'), and suggests that our knowledge is increasing ('He has given us strange creatures and occurrences' and enabled the improvement of the knowledge of alchemy).

Judging from this and other works that were almost certainly written by Andreae, he also shared Bacon's belief in the need to reform knowledge so that it might be put to pragmatic uses, such as improving the art of medicine and easing the burden of human labour. This pragmatic reform of knowledge was the chief aim of the so-called Brotherhood of the Rosy-Cross, or Rosicrucian Brotherhood. On the face of it, the Rosicrucians seem to have been a secret society devoted to religious, magical and philosophical reform.

News of this secret Brotherhood, which appeared in three published manifestos, caused a stir among intellectuals all over Europe. Some of those who sympathised with their ideas on reform tried all they could to make contact with them. Most famous of these was the French philosopher René Descartes. But he, like everybody else who tried to join the Brotherhood, failed. The truth is that the Rosicrucian Brotherhood was never anything more than another fiction devised by the author of *Christianopolis.* Andreae later admitted to writing the *Chemical Wedding of Christian Rosencreutz* (1616), third of the Rosicrucian manifestos, and it now looks as though he also wrote *Fama Fraternitatis* and *Confessio Fraternitatis* (*Confession of the Brotherhood*, 1615).

Andreae was quickly dismayed by the response to what he obviously saw as a means of testing out opinion, to see how many would be willing to subscribe to Rosicrucian reformist principles. He was disappointed, for example, by the 'commotion of imposters and swindlers' who thrived upon it and '*falsely* called themselves Brothers of the Rosicrucians' (some of whom you can occasionally still see advertising the benefits of Rosicrucian occultism in newspaper advertisements!). But equally dismaying was the response of the learned, who continued as of old to debate the pros and cons instead of doing anything practical.

Andreae tells us of these disappointments at the beginning of his *Christianopolis*, his next literary attempt to extol the virtues of a reform of learning. Much of Andreae's utopia is concerned with describing the kind of education provided for the children. This is not just a matter of discussing the subjects taught, but also of the school environment, which, instead of being like 'Bridewells or Slaughterhouses', is nothing short of paradisiacal, and the nature of the teachers, who are 'the choice of all citizens' rather than 'the dregs of human society'.

In a generally broad curriculum there is particular emphasis upon alchemy, natural history, anatomy, mathematics, astronomy and natural philosophy. Andreae takes pleasure in taking his readers on a tour through the laboratories, anatomy theatres, halls of physics and so on, which are an important part of the educational facilities. The museum of natural history holds specimens of all natural things that are beneficial or harmful to humankind, and there is a curator on hand to explain their uses and properties. Although *Christianopolis* is often concerned with other aspects of its utopian life, there is no shortage of indications that its author still believes in the important role of natural magic and religion in promoting the more pragmatic and useful learning that was the main aim of the imagined Brotherhood of the Rosy-Cross.

Bacon never explicitly mentions Rosicrucianism, but it is difficult not to see suggestive parallels between Andreae's writings and Bacon's *New Atlantis*. Certainly it was easy for would-be Rosicrucians in the succeeding generation to present Bacon's utopia as a Rosicrucian fable. In the *Holy Guide* (1662), for example, the English alchemist and astrologer John Heydon slipped explicit references to the Brotherhood into his re-telling of *The New Atlantis*.

Even without putting words into his mouth, though, Bacon often sounds like one of the brotherhood. At the beginning of the second book of the *Advancement of Learning*, for example, Bacon suggests that there should be 'a fraternity in learning and illumination', under the paternity of God, 'who is called the Father of illuminations or lights'. If the Rosicrucian Brotherhood was not a fraternity of illuminati, it was nothing.

The most prominent feature of Bacon's own utopian account is not the description of the main town, Bensalem (which for his readers with a dilettante knowledge of Hebrew would have stood for 'New Jerusalem', a name itself that resonated with millennial expectations), but a remarkable institute devoted to research and learning, called Salomon's House (named after King Solomon in the Bible, who was, of course, renowned for his great wisdom). The members of this institute are referred to as the Brethren of Salomon's House, and if that does not sound sufficiently similar to the Brotherhood of the Rosy-Cross in itself, the description of the institute – which concentrates on the various 'houses' devoted to the study of various aspects of the natural world, as well as menageries, aquaria, museums and other facilities – is highly reminiscent of Andreae's description of the facilities for learning in his utopia. There is also a similar emphasis on the need for moral rectitude and religious devotion among the brethren.

Described as 'the noblest foundation, that ever was upon the Earth, and the lantern of this kingdom', Salomon's House was dedicated to discovering the 'true nature of all things whereby God might have the more glory in the workmanship of them, and men the more fruit in the use of them' (*The New Atlantis*). It is clearly this aspect of Bacon's utopian vision that excited him most. It was, if you like, the physical embodiment of the new method of gathering and confirming knowledge of nature that Bacon tried to develop in detail in his other works.

Indeed, Bacon was so excited by his account of Salomon's House that he forgot the main purpose of *The New Atlantis*. Intended as an illustration of the kinds of benefits to society that might result from Bacon's Great Instauration, if it could be accomplished, Bacon failed to spend sufficient time outlining the nature of the ideal society at large (rather than within Salomon's House itself). Instead of presenting the reader with a tempting foresight of 'things to come', as H.G. Wells and other more recent science fiction writers have done, he focused on what he saw as the kind of research institute that might prove fruitful in generating largely unspecified benefits. As a result, *The New Atlantis* turns out to be not so much an account of a perfect society, but an account of a perfect scientific institution. It was of interest, therefore, mainly to subsequent natural philosophers and not to the more general reader.

But perhaps this is what Bacon intended after all. Writing close to what he must have known was the end of his life, Bacon knew that he could never now accomplish his Great Instauration. His only hope lay in posterity. He might therefore have tried to make *The New Atlantis* a work of general appeal, but he chose not to, preferring to address it to natural philosophers. After all, following his impeachment and his subsequent belief that he had wasted much of his time in pursuing high office, he had very good reason not to have faith in anyone except like-minded brethren in the fraternity of learning and illumination.

Seeing *The New Atlantis* principally as the literary vehicle for the depiction of a perfect research institute intended to tempt succeeding natural philosophers, rather than as a utopia intended to persuade the general reader of the benefits of science, helps to explain the lack of convincing detail about how the society of Bensalem worked. This is a common fault of utopias, but the reader might have expected something a bit more carefully thought out from an author who spent most of his life working for the Crown and who, for the latter part of his career, occupied one of the highest positions of state below the monarch.

William Rawley, the faithful secretary who arranged the posthumous publication of *Sylva Sylvarum* and *The New Atlantis* in accordance with Bacon's wishes, wrote in a preface that Bacon had originally intended to include 'a frame of Laws, or of the best state or mould of a commonwealth'. In the end, however, instead of considering such matters of state, Bacon spent the time 'collecting the Natural History' (the material in the *Sylva Sylvarum*), 'which he preferred many times before it'. It seems that by the end of his life Bacon had lost all enthusiasm for law and politics. He'd come a long way.

CHAPTER 11

WRITING PHILOSOPHY LIKE A LORD CHANCELLOR: BACON AND THE BUREAUCRACY OF SCIENCE

We've seen how the two most salient character-istics of Bacon's new philosophy of nature – its experimentalism and its pragmatic concern to extend human power by increasing our ability to exploit natural phenomena – derived chiefly from the magical and religious traditions that were important and pervasive elements in the historical background to Bacon's work. But if we look at the fine detail – the precise prescriptions that Bacon suggests as the best way of bringing about the New Philosophy that he envisages to be the outcome of his Great Instauration – there is a strong sense of something else at work influencing his whole approach. This striking and somewhat idiosyncratic aspect of his work was first recorded by his and James I's doctor, William Harvey.

Harvey was no mean experimental philosopher himself, being the discoverer of the circulation of the blood (although he didn't publish this discovery until after Bacon's death, and there is no evidence he ever told Bacon about it). But evidently he had scant respect for Bacon's recommendations about the way to improve natural philosophy. His comment, recorded by John Aubrey, sums it up perfectly: Bacon 'writes philosophy like a Lord Chancellor'. If Harvey ever said it (where Aubrey is concerned you can never be sure), he must have said it with a sneer, implying that Bacon should just stick to being Lord Chancellor and leave natural philosophy alone.

As far as Bacon was concerned, though, writing philosophy like a Lord Chancellor was what made his writings blueprints for the most powerful and profitable reform of philosophy imaginable. Only a statesman like the Lord Chancellor could set up and supervise the necessary collaborative enterprise, which was similar in many respects to other royal ordinances. Writing in the *New Organon*, Bacon insisted that 'a collection of history natural and experimental, such as I conceive it and as it ought to be, is a great, I may say a royal, work, and of much labour and expense' (I, Aphorism 111).

In case this was overlooked, he repeated it in the accompanying natural history, the *Parasceve*.

For a history of this kind, such as I conceive and shall presently describe, is a thing of very great size, and cannot be executed without great labour and expense; requiring as it does many people to help, and being (as I have said elsewhere) a kind of royal work.

When Bacon wrote to Lord Burghley in 1592, declaring that he wanted to take all knowledge for his province, he also made it clear that he would require 'commandment of more wits than a man's own'. It had to be a major collaborative effort. 'The materials on which the intellect has to work are so widely spread', he wrote in *Parasceve*, 'that one must employ factors and merchants to go everywhere in search of them and bring them in'. In the dedication of the *Great Instauration*, Bacon stopped beating about the bush.

> *I have a request to make – a request no way unworthy of your Majesty, and which especially concerns the work in hand; namely, that you who resemble Solomon in so many things … would further follow his example in taking order for the collecting and perfecting of a Natural and Experimental History true and severe (unencumbered with literature and book-learning), such as philosophy may be built upon …*

Drawing on a tradition that Solomon possessed a perfect natural history (a tradition that Bacon also employs in *The New Atlantis*, telling us that the Brothers in Solomon's House have a copy of Solomon's legendary *Natural History*), Bacon directly seeks James's cooperation in helping him to compile an equally all-encompassing natural history. What Bacon has in mind, after all ...

> *is not a way over which only one man can pass at a time ... but one in which the labours and industries of men (especially as regards the collecting of experience) may with the best effect be first distributed and then combined* (New Organon, *I, Aphorism 113).*

What Bacon hoped for, therefore, was to be put in charge of a new department of state, with numerous civil servants at his command. What he had in mind, as he wrote in the *Advancement of Learning*, would be an 'administration of learning'.

In his attempts to persuade James to establish such a new administration, Bacon did not just confine himself to flattering comparisons of James with Solomon. In a letter to James, written shortly after his succession to the English throne, he drew upon another traditional belief, that the Empire of Persia had owed its power and domination to Persian magic. His point was *not* that the Persian magicians had the power to defeat opposing armies or that they could endow superhuman powers on their kings (remember, natural magic for Bacon and his contemporaries was very much more down to earth than it is in our fantasy-laden conception of it). According to this tradition, the Persian kings owed their success to the fact that they ruled in accordance with the correct principles of government – namely, the natural, God-given, principles of government. The Persians knew these principles because they were derived from the principles of nature that underwrote the successes of Persian magic.

The education and erudition of the kings of Persia was in science which was termed by a name then of great reverence, but now degenerate and taken in ill-part: for the Persian magic, which was the secret literature of their kings, was an observation of the contemplations of nature and an application thereof to a sense politic: taking the fundamental laws of nature, with the branches and passages of them, as an original and first model, whence to take and describe a copy and imitation for government (Advancement of Learning).

In essence, this was simply another aspect of the belief in correspondences in the Great Chain of Being, which underwrote sympathetic magic. Just as God created hierarchies in nature, so there must be hierarchies among men; as there are laws of nature, governing the behaviour of natural objects, so there must be laws among men that can be seen to be based on God's principles, not on the whims of temporary rulers. This was a standard view and Bacon was reminding James of it to persuade the King of the direct political benefits that would follow if Bacon were able to discover and establish the true principles of nature. 'There is a great affinity and consent', he wrote, 'between the rules of nature, and the true rules of policy: the one being nothing else but an order in the government of the world, and the other an order in the government of an estate' (*Advancement of Learning*).

The clearest illustration of the kind of civil administration for science that Bacon desired can be seen in *The New Atlantis*. The description of Salomon's House begins with a detailing of the facilities available for studying different aspects of nature. These facilities are so extensive that right away they suggest the need for an army of researchers, or research assistants, 'novices and apprentices', as well as 'a great number of servants and attendants'. There are caves and mines dug to different depths, high towers ('the highest about half a mile in height'), salt and fresh water lakes, artificial wells and fountains, various orchards and gardens for plants, and 'parks and inclosures' for 'beasts and birds', as well as many kinds of 'great and spacious houses' to serve as laboratories and experimental institutions for numerous other kinds of investigation. Taken as a whole, Salomon's House was to be the fulfilment of a vision that Bacon had first presented to Queen Elizabeth I in an entertainment for Christmas revels in 1594; it was to be 'so furnished with mills, instruments, furnaces and vessels, as may be a palace fit for a philosopher's stone'.

Even at the highest levels within Salomon's House there was a division of labour, with different 'offices' for the fellows or brethren.

> *[W]e have twelve that sail into foreign countries, under the names of other nations (for our own we conceal), who bring us the books, and abstracts, and patterns of experiments of all other parts. These we call Merchants of Light.*
>
> *We have three that collect the experiments which are in all books. These we call Depredators.*
>
> *We have three that collect the experiments of all the mechanical arts; and also of liberal sciences; and also of practices which are not brought into arts. Those we call Mystery-men.*
>
> *We have three that try new experiments, such as themselves think good. These we call Pioneers or Miners (New Atlantis).*

The experiments gathered in these different ways are dealt with by 'Compilers', who try to interpret the significance of the experiments and establish axioms or laws of nature. 'Dowry-men or Benefactors' then try to find ways in which this new knowledge can be exploited for the 'use and practice for man's life'. Before going any further the brethren hold a general meeting of the 'whole number' to give guidance to the so-called 'Lamps', whose job it is to 'direct new experiments, of a higher light, more penetrating into nature than the former'. These experiments are carried out by the 'Inoculators'. Finally, and most important of all, there are the 'Interpreters of Nature', who are able to draw upon the earlier work to decide on a true understanding of nature and its processes.

As a final indication of Bacon's conviction about the political importance of such explorations of nature, we are also told that the brethren ...

> *have consultations [among themselves] which of the inventions and experiments which we have discovered shall be published and which not; and take all an oath of secrecy, for the concealing of those which we think fit to keep secret; though some of those we do reveal sometimes to the state, and some not (New Atlantis).*

Here then, we have servants of the Crown (establishing Salomon's House, we are told, was the pre-eminent act of its wisest king) deciding to keep secrets from the state. No wonder that even in the very act of expressing the idea Bacon falters – first telling us they keep some things secret, then telling us they do reveal to the state some of the things they've sworn to keep secret, and finally confirming that some they do not.

The bureaucratic structure of Salomon's House, with its 'Depredators', 'Compilers', 'Inoculators' and the like, may seem nothing more than a civil servant's bizarre fantasy. But, in fact, its organisation closely mirrors the structure of Bacon's *Great Instauration*.

The 'Merchants of Light', 'Depredators', 'Mysterymen' and 'Pioneers' were all engaged in part 3 of the *Great Instauration*, the gathering of natural and experimental histories on a large scale – a scale fitting, no doubt, a forest of forests of information.

The 'Compilers', on the other hand, were engaged in part 2: 'We have three [brethren] that draw the experiments of the former into titles and tables, to give the better light for the drawing of observations and axioms out of them.' The drawing up of 'tables' forms a major part of the second book of Bacon's *New Organon*, which was intended to stand as the second part of the *Great Instauration*. This was the principal statement of Bacon's new method, and applying that method was what the 'Compilers' did.

The 'Benefactors', whose task it was to put whatever was discovered to practical use, were engaged in part 5 of the *Instauration*, which was concerned not to neglect 'anything useful that may turn up by the way'.

Finally, the 'Lamps', 'Inoculators' and especially the 'Interpreters of Nature', were engaged in part 6, which, as we read in the 'Plan of the Work', 'sets forth that philosophy which by the legitimate, chaste, and severe course of inquiry which I have explained and provided is at length developed and established'.

There's no denying that the bureaucratic organisation of science outlined by Francis Bacon is just the kind of thing we might expect from a top-flight civil servant. But there's even more than this to Harvey's claim that Bacon wrote philosophy like a Lord Chancellor. The very nature of Bacon's science seems to have been determined by his bureaucratic vision. Where most contemporary natural philosophers would have seen the advance of scientific knowledge in terms of brilliant insights made during the course of trying to solve specific problems or to understand particular puzzling phenomena, Bacon saw it as the result of a 'severe course of inquiry', or even as the result of a 'machinery' designed to produce certain knowledge.

For Bacon, scientific knowledge was the kind of knowledge that could be cranked out by the machinery he devised. Students of Balliol College, Oxford, in the days when the great scholar Benjamin Jowett was their Master, used to joke that if there was something the Master of their College didn't know, it wasn't knowledge. Bacon seems to have had a similar view of what counted as knowledge and what did not. If it couldn't be produced by his machinery, it wasn't proper scientific knowledge. A famous aspect of this is Bacon's belief that his method enables everyone to participate in the advance of scientific knowledge.

> [C]ertainly if a man undertakes by steadiness of hand and power of eye to describe a straighter line or more perfect circle than any one else, he challenges a comparison of abilities; but if he only says that he with the help of a rule or pair of compasses can draw a straighter line or a more perfect circle than any one else can by eye and hand alone, he makes no great boast. And this remark, be it observed, applies not merely to this first and inceptive attempt of mine, but to all that shall take the work in hand hereafter. For my way of discovering sciences goes far to level men's wits, and leaves but little to individual excellence; because it performs everything by the surest rules and demonstrations (New Organon, I, Aphorism 122).

Bureaucracies cannot afford to rely on people of superior capacities and gifted insights: they have to be able to work even when staffed by people of ordinary or average abilities. Accordingly, Bacon's method puts all investigators on a level.

The method works like a machine capable of being operated by anyone, or rather by any number of people working together. Bacon proposes 'to establish progressive stages of certainty' starting from the evidence of the senses, 'helped and guarded by a certain process of correction'. He rejects the 'mental operation', which he sees as the usual next stage, on the grounds that it more often than not leads to error. To avoid this, he introduces the notion of a mechanical process. After all, he points out, 'if in things mechanical men had set to work with their naked hands ... just as in things intellectual they have set to work with little else than the naked forces of the understanding', they would hardly have accomplished anything. 'There remains but one course', he wrote:

> [N]amely, that the entire work of the understanding be commenced afresh, and the mind itself be from the very outset not left to take its own course, but guided at every step; and the business be done as if by machinery (Preface, New Organon).

It's by no means clear, of course, just how Bacon thought this bureaucracy of science would work in practice. The only detailed example he provides is very striking, but it fails to convince the reader that the same approach might work in other instances. Certainly, it seems fair to say that no serious attempt was ever made by Bacon's admirers to follow suit.

The system works by the compilation of 'Tables and Arrangements of Instances' of a particular phenomenon, taken out of the general natural and experimental histories. In the second book of the *New Organon*, he provides the one and only example, dealing with phenomena of heat. He suggests compiling three tables.

The first is a table of 'Essence and Presence', which lists cases where heat is always present. Anything and everything counts: the rays of the Sun, 'all flame', liquids or anything else that has been held near a fire for a time, all bodies that have been rubbed violently, quicklime sprinkled with water, animals (although in the case of insects their heat is barely perceptible), excrements of animals when fresh, various oils, all acids and so on.

To avoid jumping to conclusions about the nature of heat, we then need to compile a table that lists seemingly closely related phenomena where heat does not occur. The rays of the Moon, for example, might reasonably be expected to be hot, but they aren't. Bacon suggests using a magnifying glass to focus the Moon's rays to see if this produces heat, as it does in the case of the Sun. There are no cases of flames, or of excrements, without heat, however. Spirit of wine hardens an egg white as though it heats it, but will it melt butter or wax? By contemplating everything in the first table, we can produce a corresponding 'Table of Deviation, or of Absence in Proximity'.

Finally, to get the full measure of the natural history of the phenomenon, we need to consider relative values, instances where intensities or degrees of heat vary, and under what circumstances. The heavenly bodies are hotter when they are directly overhead, or when they are closer to the Earth or when they are close to other heavenly bodies and have a combined effect upon the Earth. Bodies with less mass are more quickly heated than bulkier masses, so corporeal matter somehow opposes heat. And so on.

Having compiled the tables, it is now possible to apply a process of elimination to reach an understanding of the phenomena in hand. We can eliminate any suggestion that heat is concerned solely with terrestrial phenomena, for example, because we know the Sun's rays are hot. But heat cannot be a celestial phenomenon either, because of terrestrial, even subterranean, fire. 'In the process of exclusion', Bacon insists, 'are laid the foundations of true induction'. But the process of understanding is not complete until we arrive at an 'Affirmative' (*New Organon*, II, Aphorism 19). Bacon is asserting here his own notion of 'induction'. Inductive logic, regarded in the scholastic Aristotelian tradition as vastly inferior to deductive logic, usually operates by 'simple enumeration': this crow is black, that

crow is black, so is the one over there and that one
... So, *all crows are black*. This is by no means a
secure way of arguing. It might seem to hold good,
but you only need to find one counter-instance to
end up with egg all over your face. All swans were
white, until black ones were discovered in Australia.

Bacon saw worse dangers for natural philosophy in
the deductive syllogisms that Aristotelians regarded
as the only guarantee of truth, and developed his
own version of induction in order to replace deduction. The deductive syllogism, as Bacon pointed
out, 'consists of propositions, propositions consist
of words, words are symbols of notions'. The danger
is, therefore, that 'if the notions themselves ...
are confused, and over-hastily abstracted from the
facts, there can be no firmness in the superstruc-
ture. Our only hope therefore lies in a true induction' (*New Organon*, I, Aphorism 14).

It's easy to see what Bacon means. The classic
syllogism is pretty straightforward: all men are
mortal; Socrates is a man; therefore Socrates is
mortal. Nobody could argue with that. But notice
there's no necessary empirical input. The syllogism
works by conforming to definitions, to agreements
about the meanings of words. It's a valid syllogism
to say: clocks that tell the correct time are supernatural; this clock tells the correct time; therefore
this clock is supernatural. If you accept the truth
of the first premise, then the argument follows.

Aristotelians would want to say: bodies that can be made to liquefy contain the element of water in their composition; iron can be made to liquefy; therefore, iron contains water in its composition.

At least that last argument is clear. But in other cases Aristotelians were confused. A standard argument against the existence of a vacuum, for example (neither Aristotelians nor Bacon believed a vacuum could exist), was that when a body moves into a new place it has to displace the body, or matter, already occupying that place. So, drop a cube of iron into water and it has to displace the water to make way for itself. Now, since vacuum isn't a corporeal body it can't be displaced by, say, an encroaching cube of iron. Therefore, the cube of iron and the vacuum would be in the same place at the same time. But two bodies can't be in the same place at the same time, scholastics triumphantly pointed out; therefore, a vacuum is impossible. The argument depends on first denying any corporeality to a vacuum, then implying it is a corporeal body. Even though Bacon didn't believe a vacuum could exist, this was the kind of confused deductive argument that would have had him knocking his head against the wall.

So, let's get back to Bacon's example of heat. Using the information in his tables, he's eliminating all the explanations of the nature of heat that won't work, but he needs an affirmative conclusion. So, we come to what Bacon calls the 'Commencement of Interpretation', or the 'First Vintage', based on the Tables of Instances. Bacon comes to the conclusion that:

Heat is a motion, expansive, restrained, and acting in its strife upon the smaller particles of bodies. But the expansion is thus modified; while it expands all ways, it has at the same time an inclination upwards. And the struggle in the particles is modified also; it is not sluggish, but hurried and with violence (New Organon, *II, Aphorism 20).*

This is very striking because it conforms pretty well to modern notions of heat. Heat is a manifestation, we would say, of increased kinetic energy of the atomic, or molecular, particles making up the hot body. Kinetic energy is energy due to movement. So, the constituent atomic particles of hot bodies are moving faster than the particles of cooler bodies. Essentially, heat is the result of motion.

Chalk one up for Bacon. But it's by no means clear that he would have had equal success with another topic and another set of tables. Besides, what we've seen is only an example taking us to the First Vintage. 'We are to proceed', Bacon tells us, 'to the other helps of the understanding in the Interpretation of Nature and true and perfect Induction' (*New Organon*, II, Aphorism 21).

Bacon then lists nine additional aids to the intellect, including 'Rectification of Induction', 'Varying the Investigation according to the nature of the subject', and 'the Ascending and Descending Scale of Axioms'. Of these nine, however, he provided details of only the first, which he called 'Prerogative Instances'. Under this heading he lists 26 different kinds of prerogative instances – by which he means instances of phenomena most likely to throw light on any subject – and provides detailed accounts of each one. After this, even a career-bureaucrat like Bacon lacked the heart to go on, and he left the *New Organon* unfinished.

With the best will in the world towards Bacon, it has to be said that the more details of his method he provides, the less appealing it becomes. It becomes more and more like a government or a legal document, detailing the qualifications of every point. The reader begins to appreciate James I's joking comment that Bacon's philosophy is like the peace of God, 'which passes all understanding'. Certainly, it doesn't look like a procedure that was ever used by any practising natural magician, natural philosopher or natural scientist, either before Bacon or since. It looks more like the kind of thing a Lord Chancellor might dream up.

CHAPTER 12

EXTENDING THE EMPIRE OF MAN? BACON, THE LAW AND MOTHER NATURE

As well as a servant of the crown, Bacon was also a lawyer. We might expect, therefore, that his way of investigating nature might also have been influenced by his method of discovery in the law courts. We don't have to look too far. One of the Prerogative Instances that he discusses in the *New Organon*, for example, is called a 'Summoning Instance'. By way of explanation, he explicitly says this term is borrowed from the courts of law, because such instances 'summon objects to appear that have not appeared before' (*New Organon*, II, Aphorism 40). In the *Advancement of Learning* Bacon talks of a 'true coincidence between commutative and distributive justice, and arithmetical proportion', noting, for example, that 'if equals be added to unequals, the wholes will be unequal', is 'an axiom as well of justice as of mathematics'.

However, the examples of legal influences on Bacon's thought that have attracted the most attention have been those concerned with interrogating nature, and even torturing her, to learn her secrets.

It has been suggested that the interrogation of witches was seen by Bacon as a model for the interrogation of nature, and that the use of mechanical devices in experimental investigations was seen as a way of extracting nature's secrets by torture, comparable to the use of the rack or other machines of torture supposedly in use in the legal process.[2] Bacon is quoted as saying:

For like as a man's disposition is never well known or proved till he be crossed, nor Proteus ever changed shapes till he was straitened and held fast, so nature exhibits herself more clearly under the trials and vexations of art [that is, artificial devices] than when left to herself (The Proficiency and Advancement of Learning).

The clear implication is that Bacon is using an analogy to the torture chamber.[3]

Recent research has refuted suggestions that Bacon relished the torture of Mother Nature. In spite of fondly cherished myths about our grisly and gruesome past, judicial torture was never used in the English common law tradition, not even for witches. During Bacon's day torture was used in special circumstances and only by order of the monarch's Privy Council. As Bacon noted, 'in the highest cases of treason, torture is used for discovery, and not for evidence'; the aim being to identify and prevent conspiracies. Torture was not used, for example, even in the case of Essex's trial for treason. What's more, it's perfectly clear that Bacon himself always regarded torture as an abuse of power and unlikely to lead in legal proceedings to a just and accurate decision about guilt or innocence.

Those readings of his works that have been made to look like recommendations of the use of torture have been shown to depend upon a misrepresentation of Bacon's notion of 'vexation', by which he meant agitation or disturbance for testing one's mettle (drawing upon analogy with a social rather than a legal kind of testing), or simply to rely upon removing the quotations from their context. When, in his *Thoughts and Conclusions*, Bacon writes that artificial devices 'have the power to conquer and subdue' nature, and 'to shake her to her foundations', for example, he is not even discussing how to discover truths. He is merely commenting on how mechanical inventions or artificial productions like gunpowder are so dramatic in their effects – so earth-shaking, or nature-shaking – that they can create an abyss between the 'haves' and the 'have nots', so that 'one man might appear a god to another'. However, this has also been presented as though the technological discoveries are used to make nature 'betray her secrets'.[4]

It is important to bear in mind the magical and religious antecedents of Bacon's thought. Repeatedly,

Bacon insists that 'Nature cannot be conquered but by obeying her' (*Thoughts and Conclusions*). 'The chain of causes cannot by any force be loosed or broken, nor can nature be commanded except by being obeyed' ('Plan of the Work'). Man is 'the servant and interpreter of Nature' (*New Organon*, I, Aphorism 1). 'Nature to be commanded must be obeyed' (*New Organon*, I, Aphorism 3), because 'all that man can do is put together or part asunder natural bodies. The rest is done by nature working within' (*New Organon*, I, Aphorism 4).

Furthermore, being steeped in the alchemical tradition as he obviously was, he shows strong traces of still believing that the secrets of nature could only be discovered by one who had proved himself worthy. In keeping with the beliefs of sympathetic magic, and the correspondences supposed to exist between different parts of the Great Chain of Being, it was always supposed that the adept could only hope to convert base metal into noble gold, if he could also rid himself of his own base nature, and ennoble himself in the alchemical process.

For Bacon, as for so many contemporary natural philosophers, the attempt to understand nature was a religious calling, being a devotion to God's *other* book, not scripture but the book of Nature. 'Study the Heaven and the Earth, the works of God Himself', Bacon wrote, 'and do so while celebrating his praises and singing hymns to your Creator' (*Refutation of Philosophies*).

Frequently, when Bacon recommended putting presuppositions and prejudices out of one's mind to understand nature, he would think of Jesus's recommendation about the best way to enter the kingdom of heaven: 'One might say that the kingdom of nature is like the kingdom of heaven, to be approached only by becoming like a little child' (*Thoughts and Conclusions*; compare with Matthew, 18, 3).

The attitude to nature revealed in comments like these do not sit easily, to say the least, with claims that he believed nature should be tortured to give up her secrets. This modern reading of Bacon is itself tied up with various other prevalent character-isations of modern science, all of which have been foisted upon him. His desire to return mankind to the prelapsarian state, before the expulsion from the Garden of Eden when they had dominion over nature, is turned into an entirely secular ambition to dominate nature and ruthlessly exploit it.

For Max Horkheimer and Theodor Adorno, two influential cultural commentators of the twentieth century, Bacon was a symbol of a modern science which was characterised as dominating nature and tyrannising mankind.[5] The clear similarities between the utopian scientific society described in *The New Atlantis* and the Rosicrucian Brotherhood dreamed up by Johann Valentin Andreae are ignored by the feminist historian Carolyn Merchant: Andreae is seen as promoting an organic and egalitarian society, while Bacon is denounced for advancing a patriarchal and capitalist society.[6] The vitalistic and alchemical theories that Bacon clearly espoused but did not promote, pending the results of his Great Instauration, are completely overlooked and Bacon is presented as the promoter of a dead, mechanical system of nature, with vigorous antihumanist implications.[7]

It is debatable, of course, whether modern science really is as black as it is painted in these scenarios, but we needn't go into that here. The point is that these are modern opinions of the nature of science and if we want to understand Bacon's *historical* significance (which is what this book is about) we cannot do so by judging him in the light of modern opinions. Most historians agree that the kind of highly secularised, anti-humanist science described in these modern critiques of science had its beginnings in the so-called Age of Reason, the Enlightenment of the eighteenth century.

For Enlightenment thinkers, Bacon was a big hero, a herald of their way of seeing the world and their view of science, but it is easy to see that already they were making Bacon in their own image. Critics of the legacy of Enlightenment science today have forged yet another image of

Bacon, as the anti-hero who embodies everything they despise about what they see as modern scientific attitudes. There is no point trying to deny the validity of this view of Bacon, because it is a view of him that is justified by his posthumous history.

I believe that Horkheimer and Adorno, Merchant and other similar writers are wrong to use Bacon the way they do, because I feel they should go back to the man himself, but I have to accept that there is an image of Bacon, forged in the Enlightenment, which can legitimately be drawn upon to illustrate the nature of post-Enlightenment science. This image of Bacon has had a life of its own, and its own history. Although it was no fault of his, Bacon was taken up by Enlightenment thinkers and turned into a hero of their materialistic and operationalist science. When Horkheimer, Adorno, Merchant, Keller and others criticise Bacon, they are criticising the symbolic Bacon of the Enlightenment. In perpetuating this view of him, therefore, modern commentators tell us more about their own attitude towards science than they do about the real, historical Bacon. If we want to know about him, we have to escape these prejudices and look to Bacon in his own time.

The same can be said of recent attempts to implicate Bacon in the domination of the natural sciences by men. Feminist historians and philosophers of science have quite rightly pointed to the lamentable treatment of women in the history of science. Women have been badly dealt with as subjects of scientific investigation, with male scientists jumping to scientifically unsupported and culturally biased conclusions about female anatomy, physiology and psychology.

Women have also been badly treated whenever they've tried to break into the male bastions of science as practitioners; there are encouraging signs now that more and more women are beginning to break into the scientific professions, even though progress is still slow. But it doesn't end there. Feminists have pointed out that the natural sciences seem to be culturally identified, by women and men alike, with masculinity. Science is a masculine pursuit, everybody seems to think, which is unsuited to the female intellect, female perceptions, and female attitudes and ambitions.

Again, Bacon has figured largely in these feminist critiques of modern science. The responsibility for science's masculine image has often been laid at his door. In what has been described as frequent and graphic use of sexual imagery, where Nature is referred to as female and is always being exposed to male gaze, her secrets penetrated as she yields to the thrust of men's arguments, Bacon is seen as raising a 'Masculine philosophy'.[8] Apart from observing that Bacon's imagery is no more sexually graphic than that of other writers at the time, all of whom habitually referred to Nature as a female, little needs to be said. Claims that Bacon spoke glibly of the rape of nature[9] have been shown to be downright false, but he did speak in terms of a 'chaste, holy and legal wedlock' between the philosopher and Nature, and in terms of binding 'her' to the philosopher's service.

Since Bacon was one of the major figures in the formation of modern science, it is almost certainly futile to try to exonerate him from its masculinisation. But it does seem clear that, unlike the other things he introduced into natural philosophy (amalgamating aspects of natural magic with natural philosophy, advocating the experimental method, and insisting upon the pragmatic usefulness of scientific knowledge), it was no deliberately conceived part of his plan to raise a so-called masculine philosophy. Furthermore, there is no male natural philosopher from the seventeenth century, or later for that matter, who does *not* play a role in the same historical process. So, although Bacon could be said to play an important role in the history of male-orientated Western science, this gives us no unique insight into Bacon, and tells us less about him than it does about the cultural development of natural philosophy.

• Chapter 13 •

Leaving It to the Next Ages

In his will, Bacon left his 'name and memory ... to men's charitable speeches, and to foreign nations, and the next ages'. As far as his planned reform of natural knowledge was concerned, this might have seemed optimistic. It was by no means obvious that Bacon's ideas were going to appeal to posterity any more than they had appealed to Queen Elizabeth I or King James I. His epitaph might well have been Harvey's dismissive comment, that he wrote philosophy like a Lord Chancellor.

But the fates had greatness in store for Francis Bacon after all, and within a few years of his death he was, according to the English diarist John Evelyn, 'celebrated as far as knowledge has any Empire' (from the preface of Evelyn's translation of Gabriel Naude's *Instructions Concerning Erecting of a Library*, 1661). Even in France, Samuel Sorbière, historian and man of letters, recognised him as 'the greatest Man for the interest of Natural Philosophy that ever was' (*A Voyage to England*, 1709). It seems fair to say that Bacon's stock rose so high at this time that his reputation has been assured ever since. In spite of numerous attacks by philosophers and scientists, insisting that Bacon's method of science has never worked in practice and is unworkable in principle, Bacon and the Baconian method keep bouncing back.

His philosophical critics have a point, though. There is a mystery about why Bacon's ideas were taken up in such a big way. There's no denying the historical significance of his insistence on the importance of the experimental method. Similarly, his claim that knowledge should be useful for the amelioration of the human condition was undeniably of the greatest moment. But Bacon was by no means the only philosopher moving in this direction. The experimental method was proving to be increasingly successful in the mathematical sciences, and in anatomical and physiological researches. Moreover, recognition of its intellectual validity and pragmatic potential rose not only with the changing economic circumstances of the Renaissance, but also with the revival of natural magic, following the Renaissance discovery of Hermetic and other ancient magical writings. So, these things needn't have been associated with Bacon in particular. Except for one thing.

It seems clear that, with regard to experimentalism and concern with practical utility, Bacon's major contribution was as a propagandist. The significance of this should not be diminished. While natural magicians and mathematicians needed no encouragement to, and no excuse for, experimenting, it was very different for natural philosophers. Philosophers need to have *philosophical* justification for what they do. Traditionally, natural philosophers did not do experiments, and to do so would have once seemed as pointless as trying to make a philosophical point by mending a pair of shoes. And not just pointless: Bacon wrote of a damaging opinion among all philosophers,

> *[t]hat the dignity of the human mind is impaired by long and close intercourse with experiments and particulars, subject to sense and bound in matter; especially as they are laborious to search, ignoble to meditate, harsh to deliver, illiberal to practise, infinite in number, and minute in subtlety* (New Organon, *I, Aphorism 83*).

After Bacon, however, philosophers could call upon his writings, and increasingly his authority, not just to see how experiment might be useful, but also to enable them to justify to their peers why they were turning to experiment. This was certainly a major factor in Bacon's subsequent success. Although there were numerous famous experimental successes (increasing all the time) to demonstrate the importance of the experimental method, there was only one source for serious discussions of the philosophical validity of experiments. All reforming natural philosophers, therefore, had to turn to Bacon.

In the early stages, another major factor in Bacon's success, in England at least, was the obvious compatibility of his programme of reform with millenarian expectations. Millenarianism continued to be a prominent part of mainstream, orthodox Anglicanism in the early seventeenth century. A key to understanding the cryptic symbolism of the Book of Revelation written by a fellow of Christ's College, Cambridge, Joseph Mede, was published in Latin in the same year as Bacon's *New Atlantis* (1627). When it was published in English (1643), it appeared with a preface that drew attention to Bacon's apocalyptic equation between the advancement of science and the time of the end. Millenarian thinkers began to recognise in Bacon's works a blueprint for the reform of learning and for utopian social planning. Accordingly, his Great Instauration received a new lease of life. In particular his ideas were promoted by a group of Puritan reformers led by the Prussian émigré scholar, Samuel Hartlib, and the Scottish theologian, John Dury.

Hartlib's circle included a number of thinkers who would later make names for themselves as new philosophers in the Baconian mould: the mathematician and inventor, William Petty; the alchemist, George Starkey; and the alchemist and leading 'new philosopher', Robert Boyle. These and lesser-known thinkers, inspired by their religious devotion to reforms, which they saw as linked to a coming millennium, easily embraced and promoted the Great Instauration. Hartlib even tried to persuade parliament to finance the building of an institute, 'designed for the execution of my Lord Verulam's New Atlantis' (from a letter dated 8 May 1654 by Samuel Hartlib to Robert Boyle). Hartlib himself, however, fared no better in securing government aid for the Great Instauration than Bacon had.

But there was more to Bacon's success than its influence on Puritan reformers. Given the changes that were already happening in European intellectual life, his advocacy of the utility of knowledge and the experimental method were bound to be successful. He can be seen here to be jumping onto a band-wagon that was already rolling. Similarly, it is hardly surprising that Bacon's deeply held Protestant values, which found their way into his conception of the Great Instauration, should have appealed to other like-minded thinkers. On the face of it, though, it is very surprising that aspects of Bacon's science that clearly derive from his bureaucratic approach to the accumulation of knowledge should have contributed to its success. Nothing in the bureaucratic details of Bacon's method looks immediately promising. Indeed, these are the very features of his method that have caused many recent philosophers of science to dismiss his ideas as completely unworkable.

The key to understanding the way these ideas have nevertheless contributed to his success is to consider them not in abstract philosophical terms but in their historical context. Philosophers can say what they like about what works or not, but the annals of human ingenuity include a wonderfully rich and complex history of exploiting even the most unpromising things.

The succeeding generation of English natural philosophers took up suggestions in the *New Organon*, and elsewhere in Bacon's writings, that investigators should avoid leaping to premature conclusions about the workings of nature. Bacon decried such conclusions as 'anticipations of nature'. What was required, he declared, was interpretation of nature, and this could only be based on all the relevant information. It seemed as though Bacon was suggesting that the information in the natural and experimental histories should be gathered indiscriminately, without any prior attempt at sifting or organising, because any organising principles would inevitably be based on presuppositions about the way things relate to one another, and the way things are. 'The understanding', he wrote in the *New Organon*, 'must not therefore be supplied with wings, but rather hung with weights, to keep it from leaping and flying' from particulars to remote axioms, or laws of nature.

The danger to true philosophy in such leaping and flying was explained in his *History of Density and Rarity*.

It would not be difficult for me to make the scattered history which I now subjoin more systematic than it is ... But two things have persuaded me not to do that. In the first place, many of the instances are of an uncertain nature and point in many directions; and [attempting to impose] accurate order thus leads ... to error. In the second place (and this is what is really at stake in my distaste for any exact method), I wish to leave what I am about open to the industry of everyone to copy. But if this collection of instances had been given the coherence of any artificial and illustrative method, many would no doubt have given up ...

In other words, as he put it in the *New Organon*, once 'anticipations of nature', which he saw as always 'rash or premature', were put forward, 'they straightaway touch the understanding and fill the imagination', and people tend to be convinced by them. Interpretations of nature, by contrast, 'being gathered here and there from very various and widely dispersed facts, cannot suddenly strike the understanding'. Nevertheless, interpretations, rather than anticipations, are what is required for sound philosophy.

What looked like a recommendation to gather information without any preconceived, and therefore inherently biased, organising principles was further reinforced by Bacon's admission in the *New Organon* that in spite of long and patient work he was still not in a position to interpret nature properly: 'I candidly confess that the natural history which I now have, whether collected from books or from my own investigations, is neither sufficiently copious nor verified with sufficient accuracy to serve the purposes of legitimate interpretation ... I have no entire or universal theory to propound', he reiterated, 'for it does not seem that the time has come for such an attempt.' Finishing his point with a suitably apocalyptic image, Bacon insisted that he was content not to 'make haste to mow down the moss or the corn in blade, but wait for the harvest in its due season' *(New Organon*, I, Aphorisms 116, 117).

Now, any philosopher of science will tell you that scientific knowledge does not advance (and never has) by the indiscriminate collection of facts. The collection of facts prior to interpretation, which Bacon seems to advocate, is a nonsense, they will tell you, because facts themselves can be discerned only within an interpretative framework. When a witness to a crime is asked by a policeman to give the facts, that witness doesn't start by describing his or her own socks, or by pointing out that there was no hot-air balloon in the sky at the time, even though these are facts about the scene.

Some philosophers, eager to test their mettle against Bacon, have pointed this out and dismissed the great man for a fool. Other philosophers, seeking to defend him, have pointed to places where Bacon seems to be aware that some kind of interpretative framework is indeed essential in compiling natural and experimental histories. Aphorism 82 of the *New Organon*, for example, rejects 'simple experience, taken as it comes' as 'a mere groping, as of men in the dark', when what is required is to 'wait for daylight, or light a candle, and then go'. In the *History of Density and Rarity*, after dismissing the imposition of an artificial order on the data, he declares: 'All the same, I shall not altogether neglect arrangement in setting out the instances ... but will put them together in such a way that they shed mutual light on each other.'

Most famously of all, Bacon offers another of his wonderful images to illustrate the true business of philosophy:

Those who have handled sciences have been either men of experiment or men of dogmas. The men of experiment are like the ant; they only collect and use; the reasoners resemble spiders, who make cobwebs out of their own substance. But the bee takes a middle course; it gathers its material from the flowers of the garden and of the field, but transforms and digests it by a power of its own (New Organon, *I, Aphorism 95).*

The true philosopher, Bacon suggests, behaves like the bee by putting the 'matter which it gathers from natural history' into the understanding 'altered and digested' (although notice he talks about gathering not from nature, but from the natural histories, which leaves open the claim that he believed the histories could be gathered in an indiscriminate way).

But both kinds of philosopher, for and against, are missing the historical point. They can argue all they like about whether Bacon was a poor philosopher, or a pretty good one, but given the contradictory evidence, no firm decision can ever be reached; it can only be a matter of opinion. If we turn away from philosophy and look to history instead, we can see straight away that the notion of indiscriminate collection of facts, with no preconceived interpretative principles, was definitely seen as a characteristically Baconian precept, *and* that (workable or not) it was immensely influential on the following generation of natural philosophers. That, if philosophers will forgive me for saying so, is a fact.

But this requires explanation. Why should this peculiar approach to the would-be advancement of knowledge have proved so popular with seventeenth-century natural philosophers?

To understand this, we have to consider changes in the age-old tradition of natural philosophy as a handmaid to religion (see Chapter 8). Before the Reformation there was a monolithic Church – the Roman Catholic Church – and it was supported by its handmaid, a carefully interpreted and, where necessary, modified Aristotelianism. After the Reformation, however, the disputes between the different factions or sects of the Church were reflected in the ways that different anti-Aristotelian natural philosophies were taken up in support of the different factions. The inevitable result was an increasing distrust among the educated of all new versions of natural philosophy, and even of traditional Aristotelianism. Where Aristotelianism once seemed to be a neutral non-partisan philosophy, whose truth meant that it could support the truth of the Roman Church, it was now rejected by many Protestants. The Roman Church was not true, they believed, and therefore Aristotelianism must be a false philosophy deliberately forged to bolster the Anti-Christian Papacy (Aristotle, after all, was a pagan, not a Christian).

Other Protestants, however, smelled an even more pernicious rat. They thought that the new philosophies were dreamed up by Catholics to confuse and distract the educated, embroiling them in futile philosophical disputes, while Jesuits and others worked on the uneducated common people to win them back to the false Church of Catholicism. Sir John Barber (King's physician and librarian of the Bodleian Library at Oxford University), for example, believed: 'It is certain this New Philosophy (as they call it) was set on foot, and has been carried on by the Arts of Rome … for the great writers of it are of the Roman religion' (he mentions by way of illustration three French philosophers, including René Descartes). Referring to religious disputes in the Netherlands and to the English Civil War, he goes on: 'What divisions this new Philosophy has caused amongst Protestants in Holland and England, cannot be unknown to any considering person' (from a letter by Barber written in 1675 to Thomas Barlow, Bishop of Lincoln).

One way or another, nobody trusted Aristotelianism any more. But the new philosophies seemed equally untrustworthy. Purporting to describe the natural world as it really was, in fact they were deliberately formed in order to support one religious faction or another, or so it seemed. Where natural philosophy was concerned, the educated were increasingly inclined to wish a plague on all their houses. For those who still believed that the study of the natural world was a noble and even a religious calling, these contemptuous attitudes were highly disturbing. It's quite clear, however, that many of the natural philosophers believed they could save the day, and restore the integrity of natural philosophy, by espousing a vigorously Baconian approach. By claiming to be good Baconians, English natural philosophers could deny that their researches into natural phenomena were intended to support any particular philosophical system, much less any supposedly affiliated religious sect. They were not putting forward any theories, still less any all-encompassing system of knowledge. They were simply engaged in collecting and establishing entirely neutral, we would say objective, 'matters of fact'.

After all, Bacon himself was explicit in the *New Organon*: 'I must request men not to suppose that after the fashion of ancient Greeks, and of certain moderns ... I wish to found a new sect in philosophy. For this is not what I am about.' Bacon's strictures against anticipations of nature could easily be made to look like condemnations of the kind of distorted, ideologically biased systems of philosophy that were being put to use by radical sectarians, Roman Catholics, or even, it was believed, atheists. Again, in the *New Organon*, Bacon had pointed out that 'in sciences founded upon opinions and dogmas, the use of anticipations is good; for in them the object is to command assent to the proposition, not to master the thing'.

Bacon seems to have put his finger on the very reason for the distrust of natural philosophy that arose during the Civil War period. Anticipations were powerful in 'the winning of assent', Bacon had suggested, but they were rash, premature and untrustworthy. In particular, 'admixture of theology' leads to 'corruption of philosophy' and 'does the greatest harm'. His system, however, of gathering facts for the natural and experimental histories, and deferring judgement about the axioms or laws of nature that were revealed by these, until certainty could be assured, until the time is right for the harvest, seemed to offer a promise of a truly reliable, unbiased and objective science of nature. The Baconians, like their newly chosen mentor, believed that the point was not to develop an 'entire or universal theory', but to sow 'in the meantime for future ages the seeds of a purer truth' (*New Organon*, I, Aphorism 116).

What we are seeing here, of course, is not really a change in how to go about doing natural philosophy. Philosophers of science are perfectly correct; indiscriminate collecting of facts without any theoretical position is unworkable. What we are seeing is the appropriation of a useful interpretation of Baconianism into the *rhetoric* of science.

Baconian rhetoric can be seen at work in the writings of every late seventeenth-century English natural philosopher who is remembered in the history of science. But it is most obvious in a book that was written to defend and promote the collective work of one of the earliest scientific research institutions, the Royal Society. Although sometimes referred to as The Royal Society of London, for the Improving of Natural Knowledge, in 1660 (when it was founded) it was the first Royal Society of any kind in Britain so it was not necessary to call it anything other than the Royal Society. This is how it is known today, having long since established itself as one of the most prestigious scientific organisations in the world.

In 1660, however, at the Restoration of the English monarchy (after the Civil War, the trial and execution of the King for, of all things, treason, and the political experiments of republicanism and the Protectorate), the newly founded Royal Society was not so prestigious. Indeed, as one of the fellows reported, it was being popularly condemned as 'a Company of atheists, Papists, dunces and utter enemies to all learning' (from a letter dated 31 January 1670 by Joseph Glanvill to Henry Oldenburg, Secretary of the Royal Society). As a society devoted to natural knowledge, it was a victim of the wide distrust of natural philosophy that had grown during the Civil War period.

Baconianism was the answer, but the educated public needed to know what Baconianism was (in the required sense, especially) and why it guaranteed the integrity of the brand of natural philosophy produced by the Royal Society. So, the Society commissioned a book of self-promotion, written under strict supervision of the leading members by Thomas Sprat, an up-and-coming young man of letters, later to become Bishop of Rochester. The book was given a simple Baconian title: *The History of the Royal Society* (1667). Furthermore, Bacon figures prominently in a famous engraved frontispiece, as the *Artium Instaurator*, Restorer of the Arts (meaning, not the fine arts, but the mechanical, mathematical, medical and magical arts; see Figure 5).

In his 'model of their whole design', Sprat tells the reader that the purpose of the fellows is 'to make faithful records of all the Works of Nature, or Art'. The information gathered will be so wide that it cannot be 'confin'd to the custody of a few', or bound in 'servitude to private interests'. It is intended to be 'not onely an Enterprise of one season … but a business of time; a steddy, a lasting, a popular, an uninterrupted Work'. In case the point is missed, he makes it explicit:

> *They have attempted to free [knowledge] from the Artifice, and Humours, and Passions of Sects; to render it an Instrument, whereby Mankind may obtain a Dominion over Things, and not onely over one another's Judgements.*

The Baconian tone is unmistakable. The members 'openly profess, not to lay the Foundation of

Figure 5: The frontispiece to Thomas Sprat's *History of the Royal Society of London, for the Improving of Natural Knowledge* (London, 1667). Bacon is depicted on the right as the *Artium Instaurator* (Restorer of the Arts). The bust is of King Charles II, described on the plinth as 'Author and Patron' of the Society; to the left is the first President of the Society, William Brouncker.

an English, Scotch, Irish, Popish, or Protestant philosophy; but a Philosophy of Mankind'. The Royal Society was an Assembly engaged 'on a design so public, and so free from suspicion of mean, or private Interest'. Only by pursuing the Baconian method can the failures of former ages be avoided, ages in which intellectual achievement had been blighted by 'Interest of Sects, the violence of Disputations ... the Religious controversies, the Dogmatical Opinions, ... and the want of a continual race of Experimenters'.

The Royal Society even lives up to Bacon's bureaucratic model of science. Describing 'Their weekly Assemblies', Sprat tells us that the material that has been collected in their natural histories is 'brought before their weekly meetings, to undergo a just and a full examination'. These meetings, according to Sprat, are characterised by 'moderation of dissenting', and 'sobriety of debating', and are not only 'free from Faction, but from the very Causes and beginnings of it'. And although the work on which they are engaged 'is vast, and almost incomprehensible, when consider'd in gross', it is made 'feasible and easie, by distributing the burden'. It is a collaborative enterprise on a large scale. Such a collaboration can work because it does not require 'a select company of refin'd Spirits', but 'may be promoted by vulgar hands'. (We tend to use the word 'vulgar' in a pejorative sense now, often meaning simply 'rude', but

Sprat was using it to mean 'common to all', or 'of the common people'.)

So, fellows of the Society 'exact no extraordinary praeparations of Learning: to have sound Senses and Truth, is with them a sufficient Qualification. Here is enough business for Minds of all sizes: ... from the most ordinary capacities, to the highest and most searching Wits'. The Society's method, therefore, like Bacon's, 'places all wits nearly on a level'. They even developed schemes that could be seen as examples of the 'machinery' for gathering information, which Bacon envisaged. Leading members of the Society drew up questionnaires, or lists of desiderata, for distribution to sea-captains and other travellers, with a view to compiling consistent information about foreign lands, their flora and fauna and so on, for inclusion in natural and experimental histories.

Even in the case of experiments, it was declared to be the 'work of the Assembly', not of the experimenter himself, to 'judge, and resolve upon the matter of Fact' – which is to say, to decide upon what fact was actually revealed by the experiment. Experiments do not, of course, speak for themselves – they have to be interpreted. Consider a bubble chamber used in high-energy physics laboratories today. Streaks of bubbles that appear in the superheated liquid kept under pressure in the chamber are said to reveal the tracks of subatomic particles.

Most of us believe this simply because that's what we are told. For all we really know the streaks might be caused randomly, like wrinkles in the skin on top of hot milk.

Conscious of this need for interpretation, and wishing to avoid suspicions among his readers that experiments were interpreted in such a way as to bamboozle or pull the wool over onlookers' eyes (don't forget, most people at this time still associated experiments with magicians, who were all too often frauds and charlatans), Sprat was eager to reassure people. He made it plain that the experimental results of the Society were not described in accordance with one person's biased view of things. They were decided by the Assembly, and what's more they were interpreted in the most matter of fact way; the issue was never about what abstract theory was confirmed in the experiment, but about 'those things, which are the plain objects of their eyes'.

Given the nature of the procedure, the experimental results of the Society established not theories but matters of fact that were 'out of all reasonable dispute'. In true Baconian fashion, the fellows did not experiment to test hypotheses, or so Sprat claims.

For whosoever has fix'd on his Cause, before he has experimented; can hardly avoid fitting his Experiment, and his Observations, to his own

Cause, which he had before imagin'd; rather than
the cause of the truth of the Experiment it self
(The History of the Royal Society).

Accordingly, the Society's fellows 'always professed to be so backward from setling of Principles, or fixing upon Doctrines'. Indeed, he even goes so far as to say 'they have wholly omitted Doctrines'. The Royal Society, therefore, pursued the slow, steady accumulation of matters of fact by gathering natural and experimental histories, and they did so without any preconceived universal theories about the underlying natures of things. What's more, in so doing they were setting on foot the method of Lord Bacon, the 'one great Man, who had the true Imagination of the whole extent of this Enterprise'.

The Royal Society was explicitly compared by some of its fellows to Salomon's House in *The New Atlantis*, and Sprat's account makes it clear that the similarity was not just based on the obvious fact that it was an institute devoted to collaborative scientific research. The fellows even wanted to claim that they were pursuing a programme and a method exactly like that described in the *Great Instauration*, and imaginatively envisaged in Salomon's House. In making this claim, it is clear that they wanted to convince their contemporaries that the scientific knowledge they were producing was not a concocted theory or system, intended to support a particular religious or political position, but was compiled simply from matters of fact, which were objective, undeniable and true.

The Royal Society went on to become extremely successful and influential, being seen as the embodiment of science in England. The fact that its actual successes rest more on the discoveries of individual innovative natural philosophers among its fellows than on its professed Baconian collaborative schemes is, as far as the success of Baconianism is concerned, neither here nor there. What mattered for the posthumous fame of Bacon was the fact that the members continued to rhetorically exploit the advantages of Baconian objectivism and factualism. Perhaps the seal was set upon this when Isaac Newton, champion of English science, defended his theory of gravity against the attacks of the German philosopher, Gottfried Wilhelm Leibniz, in entirely Baconian terms.

Leibniz was baffled and appalled by Newton's effrontery in proposing gravity as an occult force of attraction capable of acting across vast distances of empty space. For Leibniz, part of the role of the physicist, perhaps the most important part, was to *explain* the phenomena on which he based his physics. He demanded, therefore, that Newton should offer an account of how gravity works. What he would have had in mind would have been an account based on the putative continual movements of invisibly small particles capable of producing some kind of downward pressure on things near the earth – if not that, then some kind of physical, non-occult account. But for Newton, this would have meant deviating from the principles of Lord Bacon. It would mean coming up with a rash anticipation of nature, instead of waiting for the harvest in due course.

Newton was content, therefore, to uphold gravity simply as an undeniable 'matter of fact' (you can demonstrate its matter-of-factness, experimentally, by letting go of this book). 'I feign no hypotheses', Newton famously wrote in the 'General Scholium' to his *Principia Mathematica* (second edition, 1713), because 'hypotheses ... have no place in experimental philosophy' (ask any philosopher of science and they'll disagree with this remark, notwithstanding the fact it was made by the great Isaac Newton). 'In this philosophy', Newton went on, 'particular propositions are inferred from the phenomena and afterward rendered general by induction ... And to us it is enough that gravity does really exist and act according to the laws which we have explained.' For English readers at least, this was a clear endorsement by Newton of Baconian induction and a claim to be dealing only with Baconian matters of fact.

• CHAPTER 14 •

WAITING FOR THE HARVEST?

This was the ultimate endorsement. Newton's was a name with real clout, and the future of the Baconian method was now assured. When the leading thinkers of the French Enlightenment recognised the power and potential of Newton's achievement, they looked more closely at the culture from which his work had emerged. Recognising first of all Newton's older contemporaries – the experimental scientist Robert Boyle and the philosopher John Locke – they soon noted the influence on them of Francis Bacon.

When Enlightenment thinkers wrote of the precursors to their own age, which they saw as an Age of Reason, they spoke of Bacon, Descartes, Locke and Newton. The influential French thinker Voltaire saw Bacon's method of science as another example of what he discerned as a characteristically English way of avoiding controversy and dispute, by taking a middle road. In his *Thoughts and Conclusions*, Bacon had said of himself that 'he proposed a work on the interpretation of nature and on nature itself, designed to eradicate errors with the least possible offence and thus to effect a peaceable entry into the apprehensions of men'. Voltaire clearly believed that Bacon had indeed succeeded in doing this.

Voltaire's admiration for the constitutional monarchy of English politics, as opposed to the tyrannical absolutism of the French monarchy (this was in the years between the so-called Glorious Revolution of 1688 in England, which gave Parliament power over the monarch, and the French Revolution), was linked in his mind to the experimental method of the Royal Society, and therefore of Bacon. Voltaire saw England as a country where freedom of thought was cherished, and he saw this as another outcome of the Baconian rejection of the *esprit de système* (the 'spirit of philosophical systems', which he saw as enslaving French minds) and all ideological authority.

Increasingly, Lord Chancellor Bacon was seen by Enlightenment thinkers as the greatest influence on English ways of thinking, which resulted in their superior political system and their superior science. It is entirely fitting, therefore, that one of the great monuments of the French Enlightenment, the *Encyclopaedia, or Reasoned Dictionary of the Sciences, Arts, and Crafts* (1751–65), a multi-volume and would-be fully comprehensive guide to all knowledge, edited by Denis Diderot and Jean D'Alembert, but written collaboratively by many contributors, should have been explicitly presented as an embodiment of the natural and experimental histories required for the Baconian Great Instauration. For the editors of the *Encyclopaedia*, Bacon was 'the greatest, the most universal, and the most eloquent of the philosophers'.

Bacon continued to find willing champions but, as we saw in the previous chapter, arguments began to be raised against the validity of the Baconian method. It was increasingly pointed out, perfectly correctly, that the claim to accumulate data without any theories or other preconceptions was completely unworkable in practice. Such critics used history to show that no great scientist (not even Darwin, in spite of his autobiographical note, quoted at the beginning of this book) had ever used the Baconian method.

None of them, however, thought to use history to help them uncover why this unused and unusable version of the Baconian method had nevertheless made Bacon's name. They failed to note how this selective and truncated form of what Bacon really had in mind still managed to serve, not only to establish the experimental method and the concern with practical progress for the amelioration of the human condition, but also to establish the ideological neutrality and objectivity of scientific knowledge. Like the cultural commentators and feminist philosophers we looked at in Chapter 11, these philosophers of science were not attacking Bacon himself, but the image of Bacon and his method forged for their own purposes by the founding fellows of the Royal Society and the *philosophes* of the French Enlightenment.

Enlightenment intellectuals were entirely optimistic about the new science and the changes it would bring to all aspects of human living. This makes Bacon a particularly suitable hero for the Enlightenment. He too was always optimistic about his plan to reform the world of learning, and about the benefits that would accrue to mankind if knowledge of nature should be pursued for practical ends, so that human knowledge and human power met in one.

Bacon's optimism was no doubt inspired by his conviction that human dominion over nature was part of his God's original plan for mankind, and the restoration of such dominion could only help in the fulfilment of human destiny. In our secular age, human dominion over nature does not seem such an unequivocal benefit, and there is no unanimity in the belief that the fulfilment of human destiny will be a good thing, rather than an evil one. The exploitation of scientific knowledge, which once seemed to bring nothing but benefits to mankind, now seems to be fraught with danger and the potential for destruction. But perhaps we should keep our faith with science. Like Bacon, we should try to be forever optimistic in spite of the onslaught of setbacks. Perhaps, like Bacon in his *Thoughts and Conclusions*, we should note 'this further ground for comfort'.

When he reviewed the infinite expenditure of brains, time, and money on objects and pursuits which, fairly judged, are useless, Bacon was certain that a small portion of this expenditure devoted to sane and solid purposes could triumph over every obstacle

It is a final irony of Bacon's thought that his optimism even displaced his belief in induction. To learn from induction is to learn from past experience, but Bacon's optimism saw a way to avoid an inductive despair. 'The blacker the past', he wrote, 'the brighter the hope for the future' (Thoughts and Conclusions). After all, he explained, 'if men had been on the right path all those ages past and yet had got no further, what hope could there be?' As things are, we know that we just have to try to find the right path, where human knowledge and human power, and we must add, human desire, meet in one. 'Not to try', he believed, 'is a greater hazard than to fail ... Not to try is to forgo the prospect of measureless good.'

GLOSSARY

Adamic: relating to Adam, the first man, as told in Genesis, the opening book of the Bible. Usually used to refer to states of affairs and phenomena before his, and Eve's (the first woman's), fall from God's grace and subsequent expulsion from Eden. Adamic wisdom, therefore, was the wisdom supposedly held by Adam before the Fall.

Anglicanism: doctrinal system of the Protestant Church in England, which had its origins in Henry VIII's break with Catholicism but achieved its distinctive doctrine, neither Catholic nor affiliated to Continental Reformed Churches, under Elizabeth I.

Antichrist: scriptural character mentioned in the Epistles of John and the Revelation as the enemy of Christ. Figured prominently in early Protestant apocalyptic literature, sometimes identified as a specific historical person, but sometimes regarded as symbolic of any threat to the church (including, for Protestants, the Papacy).

Aphorism: intended to be a short, pithy maxim or saying, expressing general truth. Bacon chose to write his *New Organon* as a series of numbered Aphorisms. In so doing he was imitating Ancient writings, such as the 'Aphorisms' of the famous Ancient medical writer, Hippocrates (*c.* 460–377 BC). In some cases, however, Bacon's so-called aphorisms are very long and hardly count as pithy maxims. Perhaps he was driven to write in short numbered sections because the time he had available to write was limited. The aphoristic style allowed him to add to his *New Organon* in moments snatched from his work as a public servant.

Apocalypse: Greek word for 'revelation' used to denote the last book of the Bible, The Revelation of St John the Divine. Because of this book's concern with the end of the present order and the beginning of a new world, other Biblical books with the same theme came to be known as apocalyptic. Often used loosely to refer to the end of things as we know them.

Apostles: the twelve disciples of Jesus (with Matthias replacing Judas, who betrayed Jesus), seen as the original founders of the Christian Church.

Aristotelian: relating to Aristotle (384–322 BC), the highly influential Ancient Greek thinker whose natural philosophy completely dominated the curriculum in the university system from the thirteenth to the seventeenth century.

Atheist: literally one who does not believe in God, but often used loosely in the early modern period, when the word was first coined, as a term of abuse applied to a believer whose religious orthodoxy or devotion was allegedly dubious.

Baconian: relating to the methodology, or philosophy of science, of Francis Bacon. Often assumed to mean different things by different interpreters.

Calvinism: belief in the version of Protestantism set out by Jean Calvin (1509–64) of Geneva. More influential in England and Scotland than the other major version of Protestantism, Lutheranism.

Correspondences: belief, arising from notions about the Great Chain of Being and assumptions that there must be a purpose in all things created by God, that seemingly unrelated things are in fact connected by sympathies or disconnected by antipathies. The heavenly bodies must correspond in some way to things on earth, so the planets to the metals, the constellations of the Zodiac to parts of the human body; plants must have correspondences with animals, and so on.

Cosmology: sometimes used to refer to the study of the general structure and organisation of the cosmos or universe (that is, the arrangement of the heavenly bodies and Earth's relationship to them), but also used to signify a complete 'system of the world', embracing cosmology in the former sense, but also the workings of nature in general-physical, chemical, biological and so on.

Day of Judgement: the last day of the present order of the world when the dead of all the ages will be resurrected and everyone will be judged to be one of the blessed or one of the damned.

Deductive logic: system of logic based on the syllogism, a formalised argument consisting of propositions leading inexorably to a conclusion – for example, all birds have feathers; a bat does not have feathers, therefore a bat is not a bird. Favoured in the Aristotelian tradition as the required form of argument in establishing natural philosophical truths.

Deism: belief in natural theology but not in allegedly revealed theology. So, God's existence and wisdom can be inferred, the deist believes, from study of the natural world, which seems to have been designed and created by a supreme intelligence. Doctrines derived purely from supposedly Holy Scriptures, however, are rejected as superstitious beliefs derived from human, not divine, artefacts.

Demonology: study of demons or art of summoning them. Demons were reputed to be consummate natural magicians and therefore capable of performing marvellous things. They were not, however, considered to be supernatural and could not perform miracles.

Enlightenment, the: name given to the period of history covering the eighteenth and early nineteenth century. Also known as the Age of Reason, it was a period when intellectuals were highly optimistic about the unlimited power and benevolence of science.

Fall, the: the result of the first act of disobedience to God of Adam and Eve in seeking to know the difference between good and evil instead of remaining innocent, as God created them. Assumed in Judaeo-Christian tradition to have had continuing consequences for all subsequent generations of mankind.

Great Chain of Being: the belief that the whole of Creation was hierarchically structured in a single sequence. Difficulties in understanding what the precise sequence from bottom to top might be (taking in literally everything on the way) led to assumptions about correspondences between different aspects of Creation, such as between the seven planets and the seven metals, and so on.

Heretic: term used to denote someone whose religious beliefs deviate from the accepted doctrines of the accuser. For Protestants, a Catholic is a heretic, and vice versa. But Catholics and Protestants alike can find heretics also among their own ranks.

Hermetic: relating to Hermes Trismegistus, supposedly the Greek god (identified with Mercury in the Roman tradition) but regarded by Renaissance scholars as an ancient sage, roughly contemporary with Moses, who was the author of a body of magico-religious writings. These writings are now known to date from the early Christian era and to be written by several hands.

Inductive logic: procedure of inference from enumeration of experience. So successive observations of similars lead to a statement of a general principle (if all observed birds have feathers, it is stated that birds must have feathers). Generally regarded as inferior to deductive logic, but Bacon wanted to elevate it above deduction by developing a notion of eliminative enumeration, followed by a confirming instance (so, eliminating false general principles and confirming a remaining general principle in some way). He never succeeded in perfecting this notion.

Lutheranism: relating to the version of Protestantism established by Martin Luther (1483–1546), initiator of the Protestant Reformation.

Macrocosm: the greater cosmos, the world or universe as a whole.

Microcosm: the lesser cosmos, or man (always depicted as man, though signifying human), supposed to correspond as a whole, and with respect to its parts, to the macrocosm (as a whole and with respect to its parts).

Millenarianism (see also **Millennialism**): belief in a thousand-year period of blessedness, usually envisaged as taking place on Earth. By some held to follow the Second Coming of Christ, by others said to precede it, and in some cases therefore held to be already under way (supposedly having begun with some significant event in the history of the Church). A common belief among many Protestants in the early Reformation.

Millennialism: another name for millenarianism.

Natural magic: system of magic based on assumptions about the natural properties and powers invested in things by God at the Creation.

Natural philosophy: the study of nature, but not coincident with our notion of science. Based almost entirely upon Aristotelian principles, it was not investigative, but concerned merely with the speculative explanation of known physical phenomena in terms of a limited range of causes.

Neoplatonic: system of philosophy deriving from later followers of the Ancient Greek philosopher, Plato (*c.* 427–347 BC). Often taking its starting point from Plato's more mystical and religious views, it was a philosophy that proved influential upon the Early Fathers of the Christian Church. Many Christian beliefs derive therefore from non-Scriptural Neoplatonic doctrines.

Paracelsian: system of magical, chemical and associated religious philosophy, based on the teachings of the radical Swiss alchemist and healer Theophrastus von Hohenheim (1493–1541), who called himself Paracelsus.

Paramagnetic: term used to denote a body that aligns itself along the lines of force in a magnetic field.

Loosely speaking, a magnetic body, attracted to a magnet, like iron.

Pentateuch: name given to the first five books of the Bible. Supposedly written by Moses, at God's dictation.

Privy Council: before the rise to power of the House of Commons in Parliament, the institution of central government, performing all functions – executive, fiscal, legislative and judicial. Its personnel was appointed by the monarch from the clergy, peerage and commoners, according to perceived merit; even the peers were often chosen by virtue of office (as in Bacon's case), rather than birth.

Protectorate: period following the failure of English republicanism set up after the Civil Wars, when Oliver Cromwell (1599–1658) became so-called Lord Protector (1653–8).

Puritan: name given, often as a term of abuse, to more extreme Protestants in England, mostly Calvinist, who were dissatisfied with Anglicanism, which they tended to see as too close to Catholicism.

Reformation, the: period of history initiated by Martin Luther's break from the Roman Catholic Church and subsequent development of other Protestant faiths, such as Calvinism. Refers to calls for the Reform of the Church.

Renaissance, the: period of European history of remarkable changes. Characterised by great advances in the fine arts, voyages of exploration and discovery, the rise of cities and of absolute monarchies, the beginnings of banking, and other changes. Being a period when Ancient Greek and Roman writings were re-discovered on a massive scale, it was designated by scholars as a period of re-birth of Ancient wisdom.

Scholastic: refers to philosophy as it was developed and maintained within the university system throughout Medieval Europe. Effectively a distinctive and characteristic version of Aristotelianism.

Science: strictly speaking, this word did not acquire its current meaning until the nineteenth century, and in Bacon's time it simply meant 'knowledge'. It is often used here, however, in an anachronistic way to designate the new combination, brought about by Bacon and others, of natural philosophy with natural magic and other disciplines, and which formed something recognisably like our idea of science.

Second Coming, the: occasion when Jesus Christ returns in glory to judge the living and the resurrected dead of all ages, and to terminate the old world order. Coincides therefore with the Day of Judgement. Expected to be imminent according to the very early Christian Church, and various millenarian groups subsequently arising periodically (usually in times of social and political crisis).

Semi-Paracelsian: term used by Graham Rees, leading Bacon expert, to describe the speculative philosophy developed and consistently upheld, but not promoted, by Bacon. The term indicates the affiliation of Bacon's philosophy to that of Paracelsus.

Signatures: characteristic signs in things that indicated their similarity to other things, with which they were supposed to 'correspond'. Supposedly imposed on things by God as clues to their otherwise occult uses.

Star Chamber: In Bacon's day, a court of criminal jurisdiction, particularly in respect of violations of royal proclamations. So called, evidently, because of stars fashioned into the roof of the apartment in the Palace of Westminster where it met.

Superstition: a belief that is founded on allegedly false or corrupt evidence, or no evidence at all. Superstitious beliefs can be found in religion, philosophy, magic, science – anywhere, in fact, where supposedly non-superstitious beliefs can be found.

Sympathetic magic: name given to natural operations that were presumed to work as a result of natural sympathies or antipathies between different bodies or substances. Sympathies and antipathies were themselves supposed to derive from presumed corresponding places in the Great Chain of Being.

Utopian: derived originally from Sir Thomas More's *Utopia* (1516), meaning 'No Place', a description of a fictional ideal society; designates any supposedly perfect society.

FURTHER READING

There's no substitute for reading an important historical figure's own words, and this is especially true of Francis Bacon, who wrote with such wonderful style.

Bacon's complete works were edited during the Victorian period and are still available in major libraries. Based on this, there is a single-volume edition of all the natural philosophical works: *The Philosophical Works of Francis Bacon*, edited by John M. Robertson (Abingdon and New York: Routledge, 2011). And a more literary selection in *The Major Works*, edited by Brian Vickers (Oxford: Oxford University Press, 2008). So far eight of a projected fifteen volumes have appeared. Five are devoted to his philosophical works. Four of these were edited by the late Graham Rees, and include his important but demanding commentaries. There are three short but fascinating pieces by Bacon in *The Philosophy of Francis Bacon*, by Benjamin Farrington (Liverpool: Liverpool University Press, 1964). And there are numerous, easily available, editions of his most popular works.

There is a superb biography of Bacon, which remains engaging and entertaining while still providing meticulous detail: *Hostage to Fortune: The Troubled Life of Francis Bacon*, by Lisa Jardine and

Alan Stewart (London: Victor Gollancz, 1998).

For fuller treatments of Bacon's work, the following books can be recommended. *Francis Bacon*, by Perez Zagorin (Princeton: Princeton University Press, 1998), is succinct and clear and, as well as his philosophy, covers his work in morals, politics, law, history and literature. As the title suggests, Stephen Gaukroger's *Francis Bacon and the Transformation of Early-Modern Natural Philosophy* (Cambridge: Cambridge University Press, 2001) concentrates on Bacon's scientific work and provides easily the best complete survey. Still the best account of the influence of Bacon's public life on his natural philosophy, however, is *Francis Bacon, the State, and the Reform of Natural Philosophy*, by Julian Martin (Cambridge: Cambridge University Press, 1992).

For short surveys of different aspects of Bacon's work by leading scholars in the field, you can't do better than *The Cambridge Companion to Bacon*, edited by Markku Peltonen (Cambridge: Cambridge University Press, 1996).

The importance of magic in Bacon's reform of natural philosophy was first discussed in a classic study by Paolo Rossi, *Francis Bacon: From Magic to Science* (London: Routledge & Kegan Paul, 1968). For the best historical study of the theory of magic, see Stuart Clark, *Thinking with Demons: The Idea of Witchcraft in Early Modern Europe* (Oxford: Clarendon Press, 1997). On magic in the origins of modern science, see William Eamon, *Science and the Secrets of Nature: Books of Secrets in Medieval and Early Modern Culture* (Princeton: Princeton University Press, 1994).

Most of the histories of millenarian movements concentrate upon the more radical sects rather than the mainstream churches. For an entertaining example, see Christopher Hill, *The World Turned Upside Down* (Harmondsworth: Penguin, 1975). There is a vast literature on utopias, but J.C. Davis, *Utopia and the Ideal Society: A Study of English Utopian Writing, 1516–1700* (Cambridge: Cambridge University Press, 1981) is hard to beat.

For a fascinating but somewhat idiosyncratic view of Rosicrucianism, consider Frances A. Yates, *The Rosicrucian Enlightenment* (London: Routledge & Kegan Paul, 1972).

If you are interested in finding out more about feminist history of science and its claims about Bacon, the classic studies are Carolyn Merchant, *The Death of Nature: Women, Ecology and the Scientific Revolution* (New York: Harper & Row, 1980) and Evelyn Fox Keller, *Reflections on Gender and Science* (New Haven: Yale University Press, 1985). For a more introductory approach, see Margaret Wertheim, *Pythagoras' Trousers: God, Physics and the Gender Wars* (London: Fourth Estate, 1995). For an opposed view, see Nieves Matthews, *Francis Bacon: The History of a Character Assassination* (New Haven:

Yale University Press, 1996).

If you are interested in the philosophy of science and want to know more about philosophical, as opposed to historical, attempts to make sense of Bacon's reformist scheme, see Peter Urbach's *Francis Bacon's Philosophy of Science: An Account and a Reappraisal* (La Salle: Open Court, 1987), which makes the best job of it.

Bacon's initial influence in England is detailed in Charles Webster's magisterial *Great Instauration: Science, Medicine and Reform, 1626–1660* (London: Duckworth, 1975). Still the best account of his later influence is to be found in Michael Hunter's *Science and Society in Restoration England* (Cambridge: Cambridge University Press, 1981).

NOTES

1. Benjamin Farrington, *Francis Bacon, Philosopher of Industrial Science*, 1949.
2. Carolyn Merchant, *The Death of Nature: Women, Ecology and the Scientific Revolution*, New York: Harper & Row, 1980, chapter 7, pp. 170–2.
3. Evelyn Fox Keller, *Reflections on Gender and Science*, New Haven: Yale University Press, 1985, p. 36.
4. Merchant, op. cit., pp. 168, 172; Keller, op.cit., p. 36.
5. Max Horkheimer and Theodor Adorno, *Dialectic of Enlightenment*, 1942.
6. Merchant, op. cit., pp. 172–7.
7. Merchant, op. cit., pp. 180–6.
8. Merchant, op. cit., pp. 168–9, 171; Keller, op. cit., pp. 35–6, 38–40.
9. Keller, op. cit., p. 37.

ICONSCIENCE

THE ICON SCIENCE 25TH
ANNIVERSARY SERIES IS A
COLLECTION OF BOOKS ON
GROUNDBREAKING MOMENTS
IN SCIENCE HISTORY, PUBLISHED
THROUGHOUT 2017

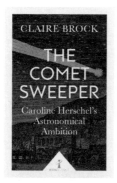

The Comet Sweeper
9781785781667

Eureka!
9781785781919

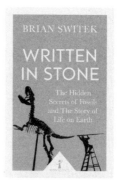

Written in Stone
9781785782015
(not available in North America)

Science and Islam
9781785782022

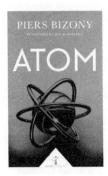

Atom
9781785782053

An Entertainment for Angels
9781785782077
(not available in North America)

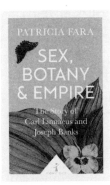

Sex, Botany and Empire
9781785782275
(not available in North America)

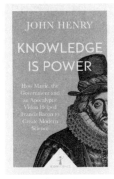

Knowledge is Power
9781785782367

Turing and the Universal Machine
9781785782381

Frank Whittle and the
Invention of the Jet
9781785782411

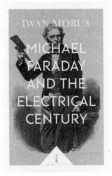

Michael Faraday and the
Electrical Century
9781785782671

Moving Heaven and Earth
9781785782695